用 英 语 介 绍 中 国

TALK ABOUT CHINESE FOOD IN ENGLISH

用英语介绍中国美食

卓燃 著

化学工业出版社

·北京·

图书在版编目（CIP）数据

用英语介绍中国美食：英、汉 / 卓燃著. -- 北京：化学工业出版社，2025.1.（2025.5重印） -- （用英语介绍中国）. -- ISBN 978-7-122-46629-7

Ⅰ. TS971.202-49

中国国家版本馆CIP数据核字第202471GH39号

责任编辑：马　骄　马小桐　　　　　　装帧设计：张　辉
责任校对：宋　玮　　　　　　　　　　版式设计：梧桐影

出版发行：化学工业出版社
　　　　　（北京市东城区青年湖南街13号　邮政编码100011）
印　　装：北京缤索印刷有限公司
787mm×1092mm　1/16　印张12　字数227千字
2025年5月北京第1版第6次印刷

购书咨询：010-64518888　　　　　售后服务：010-64518899
网　　址：http://www.cip.com.cn

凡购买本书，如有缺损质量问题，本社销售中心负责调换。

定　　价：69.90元　　　　　　　　　　　版权所有　违者必究

目录

Part 1　中国美食的**代表菜肴**

Beijing Roast Duck　北京烤鸭：北京菜的经典代表　　002

Sichuan Hot Pot　四川火锅：真正的"热辣滚烫"　　008

Kung Pao Chicken　宫保鸡丁：历史悠久的菜肴　　014

Sauteed Shrimp with Longjing Tea　龙井虾仁：西湖边上的美味　　019

Dongpo Pork　东坡肉：以苏东坡命名的经典菜肴　　025

Crab Meat Lion's Head　蟹粉狮子头：扬州的传统名菜　　031

Buddha Jumps Over the Wall　佛跳墙：浓郁醇香的享受　　037

Boiled Chicken Slices　白切鸡：粤菜中的经典　　043

Part 2　丰富多彩的**中国地方小吃**

Yangrou Paomo　羊肉泡馍：来自西北的温暖　　050

Duck Blood and Vermicelli Soup　鸭血粉丝汤：金陵的独特美味　　056

Harbin Red Sausage　哈尔滨红肠：满口都是肉香　　063

Shanghai Soup Dumpling　上海小笼包：充满鲜美的汤汁　　069

Guangxi Luosifen　广西螺蛳粉：有点"臭臭的"酸爽　　075

Guangzhou Steamed Vermicelli Rolls　广州肠粉：软糯可口的早餐　　081

Yunnan Cross-Bridge Rice Noodles

云南过桥米线：独特的米线体验　　087

Tianjin Jianbing Guozi	天津煎饼馃子：天津的经典早餐	093
Lanzhou Beef Noodles	兰州牛肉面：兰州的特色美食	099
Xinjiang Lamb Skewers	新疆羊肉串：香气四溢的烧烤	106

Part 3　历史悠久的**中国节日美食**

Mooncake　月饼：中秋节的赏月佳品	113
Zongzi　粽子：拥有2000年历史的美食	119
Yuanxiao　元宵：象征团圆的小吃	125
Dumplings　饺子：有"年味儿"的美食	131
Laba Porridge　腊八粥：腊八节的传统美食	137
Niangao　年糕：新年步步高升的象征	143

Part 4　独具特色的**中国饮食文化**

Regional Food Cultures in China　中国的地方饮食文化：风味之旅	150
Chinese Cooking Techniques　中国的烹饪技巧：中餐美味的秘诀	156
Chinese Table Manners　中国的餐桌礼仪：吃饭的文化	162
Chinese Family Gatherings　中国的家庭聚餐：重要的欢聚时刻	168
Food and Seasons　饮食与节气：顺应自然的饮食之道	174

参考答案　　　　　　　　　　　　　　　　　　　　　　　　180

Part 1
中国美食的代表菜肴

Beijing Roast Duck
北京烤鸭：北京菜的经典代表

Listening & practice 听英文原声，完成练习

▶扫码听音频◀

1. When did Beijing roast duck become well-known? all over China?

    A. Tang Dynasty

    B. Song Dynasty

    C. Qing Dynasty

2. What is used to make the duck's skin crispy before roasting?

    A. Water

    B. Vinegar

    C. Sweet syrup

3. What kind of wood is often used to roast Beijing roast duck?

    A. Pine wood

    B. Apple or cherry wood

    C. Oak wood

4. What do people usually eat with Beijing roast duck?

    A. Bread and butter

    B. Thin pancakes, green onions, cucumber, and sweet bean sauce

    C. Rice and soy sauce

5. Why is Beijing roast duck so special?

    A. It is very spicy

    B. It is eaten with rice and soy sauce

    C. It takes a lot of skill and time to make

Reading 阅读下面的文章

Beijing roast duck is one of the most famous dishes in China. It has a long history and is very delicious. Let's learn more about it!

Beijing roast duck has a history that dates back to the Northern and Southern Dynasties. It became **particularly** famous during the Ming Dynasty. The emperor loved eating roasted duck, so the palace chefs worked hard to perfect the **recipe**. They made it into a **luxurious** and special treat. Over time, the dish became more popular, and by the Qing Dynasty, it was well-known all over China. Today, everyone can enjoy Beijing roast duck, not just the royals.

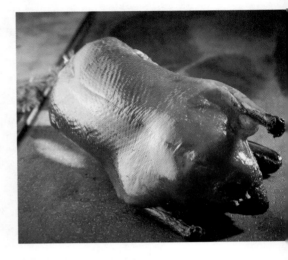

Beijing roast duck is not just food; it is part of Chinese culture. People often eat it during special **celebrations** like the Spring Festival, weddings, and birthdays. It shows the **importance** of sharing food with family and friends. Eating Beijing roast duck together brings people closer and creates happy memories.

Making Beijing roast duck is a special **process**. First, the duck is cleaned and filled with air to make the skin **crispy**. Then, it is hung to dry and coated with a sweet syrup. After that, it is roasted in an oven at a high temperature. The cook often uses fruitwood, like apple or cherry wood, which gives the duck a **unique** and delicious flavour. After roasting, the duck is **sliced** into thin pieces. The skin is very crispy, and the meat is **tender**. It takes a lot of skill and time to make Beijing roast duck, which is why it is so special.

Beijing roast duck tastes amazing! The crispy skin and tender meat are perfect together. People usually eat it with thin **pancakes**, green onions, **cucumber**, and a sweet bean sauce. You put everything in the pancake, roll it up, and enjoy! The combination of flavours and **textures** makes each bite delicious. The sweet sauce, fresh vegetables, and savory duck are a perfect match.

Vocabulary and phrases 词汇和短语

particularly [pəˈtɪkjələli] 副 特别；尤其	recipe [ˈresəpi] 名 食谱
luxurious [lʌɡˈʒʊəriəs] 形 奢侈的；豪华的	celebration [ˌselɪˈbreɪʃn] 名 典礼；庆贺
importance [ɪmˈpɔːtns] 名 重要；重要性	process [ˈprəʊses] 名 流程；过程
crispy [ˈkrɪspi] 形 脆的	unique [juˈniːk] 形 独特的；独一无二的
slice [slaɪs] 动 切成（薄）片	tender [ˈtendə(r)] 形 嫩的
pancake [ˈpænkeɪk] 名 薄煎饼	cucumber [ˈkjuːkʌmbə(r)] 名 黄瓜
texture [ˈtekstʃə(r)] 名 质地；口感	

Practice 请选择合适的词填在下方的横线上

> recipe　　process　　crispy　　unique　　texture

1. Did you know that the _____ of preparing Beijing roast duck starts with selecting the perfect duck?

2. One of the things that makes Beijing roast duck so special is its _____ cooking method.

3. Beijing roast duck is famous for its _____ skin and juicy meat.

4. I learned a fun _____ for Beijing roast duck at school today!

5. The chef at our restaurant takes great care to ensure that the _____ of the Beijing roast duck is just right.

Talking practice 情景对话模拟练习

吴迪刚到北京,他很想知道这里有哪些美食,让我们练习这段对话吧。

Beijing is the capital of China. What famous food do they have there?
北京是中国的首都,这里有什么代表性的美食吗?
Wu Di 吴迪

Daniel 丹尼尔
One of the most famous dishes is Beijing roast duck. Have you heard of it?
最有名的菜之一是北京烤鸭。你听说过吗?

Yes, I have. Can you tell me more about it?
听说过。你能告诉我更多吗?
Wu Di 吴迪

Daniel 丹尼尔
Beijing roast duck has a long history, dating back to the Northern and Southern Dynasties. It became really famous during the Ming Dynasty.
北京烤鸭有很悠久的历史,可以追溯到南北朝。它在明代变得非常有名。

How is it made?
它是怎么做的?
Wu Di 吴迪

Daniel 丹尼尔
First, the duck is cleaned and coated with a sweet syrup. Then it's roasted with fruitwood for a unique flavour.
首先,鸭子会被清洗干净并刷上一层糖浆。然后用果木烤,给它独特的味道。

What do you eat it with?
它搭配什么吃?
Wu Di 吴迪

Part 1 中国美食的代表菜肴

Daniel 丹尼尔

> We eat it with thin pancakes, green onions, cucumber, and a sweet bean sauce. You roll everything up in the pancake.
> 我们用薄煎饼、葱、黄瓜和甜面酱一起吃。你把所有东西卷在煎饼里。

> That sounds delicious! I can't wait to try it.
> 那听起来很好吃！我迫不及待想试试了。

Wu Di 吴迪

Funny facts 关于北京烤鸭的有趣事实和短语

烤鸭店：全聚德是北京最著名的烤鸭店之一，它成立于1864年，以其独特的烤制方法和口味而闻名。

鸭种选择：制作北京烤鸭通常选用北京白鸭，这种鸭子体型适中，脂肪分布均匀，肉质细嫩。

片鸭技艺：片鸭是一门技艺，讲究将整只鸭子片成均匀的薄片，通常每只鸭子可以片成百片左右。

- crispy skin – 酥脆皮
- tender meat – 嫩肉
- sweet bean sauce – 甜面酱
- green onion – 大葱
- cucumber strips – 黄瓜条
- slicing technique – 片鸭技艺

Writing practice 写作小练习

根据我们这一节所学到的内容，写出以下句子的英文。

1. 我喜欢吃烤鸭，它的皮脆脆的，肉嫩嫩的。

   _____

2. 北京烤鸭有很悠久的历史。

   _____

3. 北京烤鸭是北京最有名的特色菜之一。

   _____

Reference translation 参考译文

北京烤鸭是中国最著名的菜之一。它历史悠久,非常美味。让我们来了解一下它吧!

北京烤鸭最早起源于南北朝时期,明朝时期更是声名鹊起。当时皇帝酷爱享用烤鸭,于是御厨们不遗余力地完善烹饪方法,将其打造成一种奢华且独特的佳肴。随着时间的推移,这道菜愈发受到人们的喜爱,到了清朝,它已在中国家喻户晓。如今,北京烤鸭已不再是皇家的专属,而是人人都可以品尝的美味。

北京烤鸭不仅仅是一道菜,更是中华文化的一部分。在诸如春节、婚礼和生日等特殊庆典上,人们常常会享用它。这体现了与家人和朋友分享食物的重要性。共同品尝北京烤鸭拉近了人们之间的距离,也创造美好的回忆。

制作北京烤鸭是一项独特的工艺。首先,清洗鸭子并打气,为的是让鸭皮烤出来酥脆。然后,将鸭子挂起来风干,并涂上一层甜糖浆。接下来,将鸭子放进高温烤炉中烤制。厨师通常使用果木,如苹果木或樱桃木,这会给鸭子增添一种独特而美味的风味。烤好后,将鸭子切成薄片。鸭皮非常酥脆,鸭肉则十分嫩滑。制作北京烤鸭需要精湛的技巧和大量的时间,这也是它如此特别的原因。

北京烤鸭的味道棒极了!酥脆的鸭皮和嫩滑的鸭肉完美结合。人们通常用薄煎饼、葱、黄瓜和甜面酱与烤鸭一同享用。将所有食材放入薄饼中,卷起来,细细品味。这种味道和口感的组合使每一口都美味无比。甜甜的酱汁、新鲜的蔬菜和美味的鸭肉相得益彰,堪称完美搭配。

Sichuan Hot Pot
四川火锅：真正的"热辣滚烫"

Listening & practice 听英文原声，完成练习

▶扫码听音频◀

1. What kind of flavours is Sichuan food known for?
   A. Sweet and mild
   B. Sour and bitter
   C. Numbing and spicy

2. What makes the soup in Sichuan Hot Pot special?
   A. It has chili peppers and peppercorns
   B. It has fruits
   C. It has milk

3. What do people do when they eat Sichuan Hot Pot?
   A. They cook their own food in the pot
   B. They eat only rice
   C. They eat only fruit

4. What do Sichuan peppercorns make you feel?
   A. A cold feeling
   B. A heavy feeling
   C. A numbing feeling

5. Why is eating Sichuan Hot Pot fun?
   A. It is quick
   B. It brings people together
   C. It is cheap

Reading 阅读下面的文章

Have you ever tried something so spicy that it made your mouth **tingle**? That's what happens when you eat Sichuan Hot Pot! This famous dish from China is not only delicious but also a lot of fun to eat. Let's **dive into** the world of Sichuan Hot Pot and **discover** why it's so special.

Sichuan Hot Pot comes from Sichuan Province, a place known for its numbing and spicy **flavours**. The people in Sichuan love to add lots of **chili peppers** and Sichuan peppercorns to their food. Hot pot has been a favourite dish there for many years and has now spread all over China and even to other countries.

One of the best things about Sichuan Hot Pot is that it brings people together. When you eat hot pot, you **gather** around a big pot of **bubbling**, spicy broth with your family and friends. Everyone gets to cook their own food in the pot. You can choose from **a variety of** ingredients like meat, vegetables, tofu, and noodles. It's so much fun to pick what you want to eat and watch it cook right **in front of** you.

The broth in Sichuan Hot Pot is what makes it so special. It's made with lots of spices, including chili peppers and Sichuan peppercorns. These spices make the broth very hot and give it a unique numbing **sensation**. Some people might find it too spicy, but if you enjoy a little heat, you will love it!

Once your food is cooked, you use chopsticks to take it out of the pot and dip it in a tasty sauce. The sauce can be made from **ingredients** like soy sauce, garlic, sesame oil, and other delicious seasonings. Every bite is full of exciting flavours.

Eating Sichuan Hot Pot is a wonderful **experience**. It's not just about the tasty food but also about spending time with people you care about. It's a time to talk, laugh, and enjoy a meal together.

Vocabulary and phrases 词汇和短语

tingle ['tɪŋgl] 名 刺痛；刺激	dive into 短 投入……之中
discover [dɪ'skʌvə(r)] 动 发现；发觉	flavour ['fleɪvə] 名 风味；滋味；特色
chili pepper 辣椒	gather ['gæðə(r)] 动 聚集
bubbling ['bʌblɪŋ] 动 起泡	a variety of 短 种种；多种多样
in front of 短 在……前面	sensation [sen'seɪʃn] 名 感觉；知觉
ingredient [ɪn'griːdiənt] 名 原料；配料	experience [ɪk'spɪəriəns] 名 经验；经历

Practice 请选择合适的词填在下方的横线上

> flavour   chili peppers   gather   a variety of   experience

1. Sichuan Hot Pot has a spicy _____ that makes it tasty.

2. We often _____ around the table to enjoy a delicious Sichuan Hot Pot together.

3. Having Sichuan Hot Pot is a fun and exciting _____ for me.

4. I like to add extra _____ to my Sichuan Hot Pot to make it even spicier.

5. Sichuan Hot Pot has _____ delicious ingredients to choose from.

四川火锅：真正的"热辣滚烫"

Talking practice 情景对话模拟练习

吴迪向同学丹尼尔推荐了四川火锅，让我们跟着这段对话练习吧。

Wu Di 吴迪

Hey, Daniel, can you eat spicy food?
嘿，丹尼尔，你能吃辣吗？

Daniel 丹尼尔

Yes, I can! Why do you ask?
能啊！你为什么问？

Wu Di 吴迪

Have you tried Sichuan Hot Pot? It's very spicy and delicious.
你尝过四川火锅吗？它非常辣，非常好吃。

Daniel 丹尼尔

No, I haven't. What's special about it?
没有，我还没试过。它有什么特别的？

Wu Di 吴迪

Sichuan Hot Pot has a spicy broth made with chili peppers and Sichuan peppercorns. It makes your mouth tingle!
四川火锅的汤底是用辣椒和花椒做的，吃了会让嘴巴发麻！

Daniel 丹尼尔

That sounds exciting! How do you eat it?
听起来很刺激！怎么吃呢？

Wu Di 吴迪

You cook your own food in the pot, like meat, vegetables, tofu, and noodles. Then you dip it in a tasty sauce.
你把肉、蔬菜、豆腐和面条放在锅里煮，然后蘸上美味的酱料。

Daniel 丹尼尔

It sounds like a fun way to eat with friends and family.
听起来和朋友家人一起吃很有趣。

Yes, it's a great way to spend time together and enjoy delicious food.
是的,这是一种很好的方式,可以一起享受美食,共度时光。

Wu Di 吴迪

Funny facts　关于四川火锅的有趣事实和短语

九宫格:传统的四川火锅锅底常见九宫格设计,将锅分为九个小格子,方便不同食材分开涮煮。

食材丰富:四川火锅的食材丰富多样,包括牛肉、羊肉、鸭肠、毛肚、黄喉、虾滑、豆腐、蔬菜等。

区域差异:不同地区的火锅有各自的特色,如成都火锅偏麻辣鲜香,重庆火锅则更辣更麻。

---

broth base – 锅底
double-flavour pot – 鸳鸯锅
beef tripes – 毛肚
duck intestines – 鸭肠

chili peppers – 辣椒
Sichuan peppercorns – 花椒
dipping sauce – 蘸料

Writing practice　写作小练习

根据我们这一节所学到的内容,写出以下句子的英文。

1. 四川人喜欢吃火锅,并且会放很多辛辣的辣椒。

   _____

2. 你可以在四川火锅中煮很多种食物,比如肉、蔬菜、豆腐和面条。

   _____

3. 与家人和朋友一起吃四川火锅真的很有趣!

   _____

四川火锅：真正的"热辣滚烫"

Reference translation 参考译文

你是否曾尝过某种如此辛辣的美食，让你的舌尖忍不住颤动？这就是吃四川火锅时的感觉！这道来自中国的著名菜肴不仅美味，而且吃起来非常有趣。让我们一起探索四川火锅的世界，探寻其独特之处。

四川火锅起源于以麻辣著称的四川省。四川人喜欢在食物中加入大量的辣椒和花椒。火锅已经在四川流行了很多年，如今已风靡全国，甚至传至国外。

四川火锅最棒的地方之一就是它能让人们聚在一起。吃火锅时，你和家人朋友围坐在一个装满沸腾麻辣汤底的大锅旁。每个人都可以在锅里煮自己的食物。你可以选择各种各样的食材，如肉类、蔬菜、豆腐和面条。挑选你想吃的食物，看着它在你面前煮熟，是一件非常有趣的事情。

四川火锅的汤底是它如此特别的原因。汤底是用许多香料制成的，包括辣椒和花椒。这些香料让汤底非常辣，并且有一种独特的麻感。有些人可能觉得太辣了，但如果你喜欢吃辣的食物，你一定会爱上它的！

当食物煮熟后，你用筷子把它从锅里捞出来，蘸上美味的酱汁。酱汁可以用酱油、大蒜、芝麻油和其他美味的调料制成。每一口都充满了令人兴奋的风味。

吃四川火锅是一种美妙的体验。不仅仅是因为美味的食物，更是因为能与关心的人共度时光。这是一个谈天说地、欢笑和一起享用美食的时刻。

Kung Pao Chicken
宫保鸡丁：历史悠久的菜肴

Listening & practice 听英文原声，完成练习

▶扫码听音频◀

1. **Where does Kung Pao Chicken come from?**
   A. Guangdong Province
   B. Sichuan Province
   C. Hunan Province

2. **Who is Kung Pao Chicken named after?**
   A. A famous chef
   B. A Qing Dynasty official
   C. A Chinese emperor

3. **What are the main ingredients in Kung Pao Chicken?**
   A. Chicken, peanuts, and vegetables
   B. Beef, peanuts, and potatoes
   C. Fish, almonds, and carrots

4. **What makes the sauce in Kung Pao Chicken special?**
   A. It is made from chocolate and honey
   B. It is sweet, sour, and spicy
   C. It is salty and bitter

5. **What is usually served with Kung Pao Chicken to balance the spicy flavour?**
   A. Noodles
   B. Bread
   C. Steamed rice

Reading 阅读下面的文章

Imagine a dish that is sweet, sour, spicy, and crunchy all at the same time. That's Kung Pao Chicken! This popular Chinese dish is loved by many people around the world. Let's find out more about it.

Kung Pao Chicken is a famous dish from Sichuan Province in China. A popular story says that it was created by a Qing Dynasty official named Ding Baozhen. The dish was named after him and has been enjoyed for many years.

What makes Kung Pao Chicken special is its unique combination of flavours. The dish is made with chicken, peanuts, and vegetables. These ingredients are stir-fried together with a special sauce that is sweet, sour, and spicy. The sauce is made from soy sauce, vinegar, sugar, and chili peppers. This mix of flavours makes every bite exciting and delicious.

Cooking Kung Pao Chicken is a fun process. First, the chicken is cut into small pieces and marinated. Then, it is stir-fried in a hot pan with some oil. After the chicken is cooked, peanuts and vegetables are added. Finally, the special sauce is poured in, and everything is mixed together. The dish is ready in just a few minutes!

Kung Pao Chicken is not only tasty but also a healthy dish. It has lots of protein from the chicken and peanuts, and vitamins from the vegetables. It is usually served with steamed rice, which helps balance the spicy flavour.

Eating Kung Pao Chicken is a delightful experience. The different flavours and textures make it a favourite for both kids and adults. It's a great dish to enjoy with family and friends.

Vocabulary and phrases 词汇和短语

imagine [ɪˈmædʒɪn] 动 想象	crunchy [ˈkrʌntʃi] 形 松脆的
create [kriˈeɪt] 动 创造	combination [ˌkɒmbɪˈneɪʃn] 名 组合
vinegar [ˈvɪnɪɡə(r)] 名 醋	marinate [ˈmærɪneɪt] 动 浸泡

tasty ['teɪsti] 形 美味的；可口的   healthy ['helθi] 形 健康的
protein ['prəʊtiːn] 名 蛋白质   vitamin ['vɪtəmɪn] 名 维生素
steamed [stiːmd] 形 蒸熟的

Practice 请选择合适的词填在下方的横线上

> imagine　create　tasty　healthy　protein

1. Every bite of Kung Pao Chicken is full of _____ flavours.
2. Kung Pao Chicken is a great source of _____, which helps our bodies grow and repair.
3. I can _____ the aroma (香气) of Kung Pao Chicken filling up the whole kitchen as it cooks.
4. Kung Pao Chicken can be a _____ dish if you use lean chicken and plenty of vegetables.
5. My mom is a great cook, and she can _____ delicious dish every time.

Talking practice 情景对话模拟练习

盖文和徐丽正在兴致勃勃地聊起宫保鸡丁这道菜，让我们跟着这段对话练习一下吧。

Gavin, which Chinese dish impresses you the most?
盖文，你对哪道中国菜印象最深刻？

Xu Li 徐丽

Gavin 盖文

I had Kung Pao Chicken, and it was so delicious. This dish is famous worldwide. Where is it from in China?
我吃过宫保鸡丁，它真是太好吃了，这道菜在世界上都很有名。它是中国哪里的菜呢？

宫保鸡丁：历史悠久的菜肴

Kung Pao Chicken is from Sichuan Province. It has a mix of sweet, sour, and spicy flavours.
宫保鸡丁来自四川省。它有甜、酸和辣的混合味道。
— Xu Li 徐丽

Who created this dish?
这道菜是谁创造的？
— Gavin 盖文

A Qing Dynasty official named Ding Baozhen, created Kung Pao Chicken. The dish is named after him.
一个叫丁宝桢的清朝官员创制了宫保鸡丁。这道菜以他的官职命名。
— Xu Li 徐丽

How do you cook it?
怎么做这道菜？
— Gavin 盖文

First, marinate the chicken, then stir-fry it with peanuts and vegetables. Finally, add the special sauce and mix everything together.
首先，腌制鸡肉，然后和花生、蔬菜一起炒。最后加入特别酱汁，把所有材料混合在一起。
— Xu Li 徐丽

That sounds amazing! I can't wait to try making it myself.
那听起来很棒！我迫不及待想自己试着做了。
— Gavin 盖文

Funny facts — 关于宫保鸡丁的有趣事实和短语

国际影响：宫保鸡丁不仅在中国广受欢迎，还在全球各地中餐馆中流行，深受各国食客喜爱。

酸甜口感：宫保鸡丁以其酸甜口感著称，酸来自醋，甜来自糖，形成了独特的风味平衡。

传统手艺：制作宫保鸡丁时，鸡肉先用淀粉和蛋清腌制，这样可以使肉质更嫩滑。

diced chicken – 鸡丁	**egg white** – 蛋清
dried chili peppers – 干辣椒	**savoury and spicy** – 咸辣
soy sauce – 酱油	**sour and sweet** – 酸甜

Writing practice 写作小练习

根据我们这一节所学到的内容，写出以下句子的英文。

1. 在宫保鸡丁中，你可以找到鸡肉、花生和色彩鲜艳的蔬菜。

2. 宫保鸡丁是四川著名的美食！

3. 配着米饭吃宫保鸡丁是一种享受美食的美妙方式。

Reference translation 参考译文

想象一下，有一道菜同时融合了甜、酸、辣和脆，那就是宫保鸡丁！这道受欢迎的中国菜在世界各地都很受喜爱。让我们来了解一下它吧。

宫保鸡丁是中国四川省的一道名菜。一个广为传颂的说法是，它是清朝一位名叫丁宝桢的官员创造的，这道菜就是以他的官职命名的。多年来，这道菜一直备受喜爱。

宫保鸡丁之所以特别，是因为它独特的味道组合。这道菜是用鸡肉、花生和蔬菜制成的。这些食材与一种特别的酱汁一起翻炒，这种酱汁融合了甜、酸和辣的口味。酱汁是用酱油、醋、糖和辣椒制成的。这种味道的组合让每一口都充满了刺激和美味。

烹饪宫保鸡丁是一次愉快的体验。首先，把鸡肉切成小块并腌制。然后，在热油锅中翻炒鸡肉。鸡肉煮熟后，加入花生和蔬菜。最后，倒入特别的酱汁，把所有东西混合在一起。这道菜只需几分钟就能做好！

宫保鸡丁不仅美味，而且是一道健康的菜。它有来自鸡肉和花生的丰富蛋白质，还有来自蔬菜的维生素。它通常与米饭一起食用，这样可以平衡辣味。

吃宫保鸡丁是一种愉快的体验。不同的味道和口感让它成为孩子和大人都喜爱的菜肴。它是一道很适合与家人和朋友一起享用的美食。

Sauteed Shrimp with Longjing Tea
龙井虾仁：西湖边上的美味

Listening & practice 听英文原声，完成练习

▶扫码听音频◀

1. What famous dish can you taste in Hangzhou?
    A. Kung Pao Chicken
    B. Sauteed Shrimp with Longjing Tea
    C. Sweet and Sour Pork

2. What ingredient gives Sauteed Shrimp with Longjing Tea its unique flavour?
    A. Soy sauce
    B. Garlic
    C. Longjing tea leaves

3. Which emperor is said to have favoured Sauteed Shrimp with Longjing Tea?
    A. Emperor Kangxi
    B. Emperor Qianlong
    C. Emperor Yongzheng

4. What is the first step in making Sauteed Shrimp with Longjing Tea?
    A. Boiling the tea leaves
    B. Cleaning and peeling the shrimp
    C. Frying the shrimp

5. Why is Sauteed Shrimp with Longjing Tea considered healthy?

　　A. It is rich in protein and antioxidants

　　B. It is deep-fried

　　C. It has a lot of sugar

Reading 阅读下面的文章

　　If you travel to the **picturesque** city of Hangzhou, besides enjoying the beautiful scenery of West Lake, you can also taste the famous Hangzhou **cuisine**. One of the most **renowned** dishes is Sauteed **Shrimp** with Longjing Tea. Let's find out why Sauteed Shrimp with Longjing Tea is so unique and delicious.

　　Sauteed Shrimp with Longjing Tea gets its name from the famous Longjing tea, a very special kind of green tea. This tea is grown in the hills around Hangzhou and is known for its fresh, sweet flavour. The tea leaves are used to cook the shrimp, giving the dish a light and **fragrant** taste.

　　Sauteed Shrimp with Longjing Tea has been enjoyed for over a hundred years and is said to have been **favoured** by Emperor Qianlong of the Qing Dynasty. The combination of Longjing tea and shrimp is a perfect example of how Hangzhou cuisine uses local ingredients to create unique flavours. This dish **showcases** the **essence** of Hangzhou's **culinary** traditions by blending the subtle aroma of the tea with the sweetness of the shrimp.

　　Making Sauteed Shrimp with Longjing Tea is a simple but special process. First, the shrimp are cleaned and peeled. Then, the tea leaves are soaked in hot water to release their flavour. The shrimp are quickly stir-fried with ginger and a little bit of salt. Finally, the tea leaves and tea water are added to the pan. The shrimp **absorb** the tea's wonderful flavour, making them taste fresh and delicious.

　　What makes Sauteed Shrimp with Longjing Tea special? Sauteed Shrimp with Longjing Tea is not only tasty but also healthy. Shrimp are a good source of protein, and Longjing tea is full of **antioxidants**. The shrimp is tender and flavourful, with a light tea fragrance. This makes the dish both nutritious and delicious. Everyone loves it after just one bite.

Vocabulary and phrases 词汇和短语

picturesque [ˌpɪktʃəˈresk]
形 如画的；生动的

renowned [rɪˈnaʊnd] 形 有名的；有声誉的

fragrant [ˈfreɪɡrənt] 形 芬香的；馥郁的

showcase [ˈʃəʊkeɪs] 动 使展示；陈列

culinary [ˈkʌlɪnəri] 形 厨房的；烹调的

antioxidant [ˌæntiˈɒksɪdənt]
名 抗氧化剂

cuisine [kwɪˈziːn] 名 佳肴

shrimp [ʃrɪmp] 名 虾

favour [ˈfeɪvə] 动 偏爱；喜爱

essence [ˈesns] 名 精髓；本质

absorb [əbˈsɔːb] 动 吸收

nutritious [njuˈtrɪʃəs]
形 有营养的；滋养的

Practice 请选择合适的词填在下方的横线上

cuisine   shrimp   favour   showcases   nutritious

1. Sauteed Shrimp with Longjing Tea is a perfect dish for _____ lovers like me!

2. My friend and I both _____ Sauteed Shrimp with Longjing Tea so much.

3. My friend is a great cook, and she always _____ her talent with Sauteed Shrimp with Longjing Tea.

4. Sauteed Shrimp with Longjing Tea is a delicious part of Chinese _____.

5. Eating _____ foods like Sauteed Shrimp with Longjing Tea helps me stay strong and healthy.

Talking practice　情景对话模拟练习

在杭州，郑凯向利奥推荐了龙井虾仁这道菜，你也跟着练习这段对话吧。

Leo, have you ever tried Sauteed Shrimp with Longjing Tea?
利奥，你吃过龙井虾仁吗？

Zheng Kai 郑凯

Leo 利奥

Sauteed Shrimp with Longjing Tea? What kind of dish is it?
龙井虾仁？这是什么菜呢？

It's a famous dish from Hangzhou, made with shrimp and Longjing tea leaves.
这是杭州的名菜，是用虾和龙井茶做的。

Zheng Kai 郑凯

Leo 利奥

How do they make it?
它是怎么做的？

First, the shrimp are cleaned and peeled. Then, the tea leaves are soaked in hot water. The shrimp are stir-fried with ginger and salt, and finally, the tea leaves and tea water are added.
首先，把虾清洗干净并剥壳。然后，把茶叶用热水泡开。虾用姜和盐炒，再加上茶叶和茶水。

Zheng Kai 郑凯

Leo 利奥

That sounds delicious! Does it have a special taste?
听起来很好吃！它有什么特别的味道吗？

龙井虾仁：西湖边上的美味

> Yes, the shrimp absorbs the fresh, sweet flavour of the tea, making it very light and fragrant. Would you want to give it a try?
> 有的，虾吸收了茶的清香甜味，非常清淡和芳香。你想不想试一试？

Zheng Kai 郑凯

Leo 利奥

> Yes! I would love to try it.
> 是的。我很想尝尝。

Funny facts 关于龙井虾仁的有趣事实和短语

名厨推荐：许多名厨都推荐龙井虾仁，认为其不仅代表了杭州菜的精髓，也展示了中式烹饪的精妙技艺。

茶文化结合：这道菜将中国茶文化与美食巧妙结合，是茶宴中不可或缺的一道佳肴。

轻炒手艺：龙井虾仁的烹饪过程中，虾仁需用低温慢炒，避免过熟，保持虾仁的鲜嫩口感。

Longjing tea – 龙井茶	tender texture – 嫩滑口感
light stir-fry – 轻炒	healthy cuisine – 健康菜肴
marinated shrimp – 腌制虾仁	classic recipe – 经典食谱

Writing practice 写作小练习

根据我们这一节所学到的内容，写出以下句子的英文。

1. 龙井虾仁不仅美味，而且也非常健康。

   _____

2. 它是由美味的虾仁和龙井茶做成的。

   _____

3. 龙井虾仁是一道特别的中国菜。

   _____

Reference translation 参考译文

如果你到风景如画的杭州去旅行，除了浏览西湖的美景以外，还可以品尝一下著名的杭州菜，其中最为有名的菜肴当属龙井虾仁。让我们一起来探寻龙井虾仁为何如此独特与美味。

龙井虾仁的名字来源于著名的龙井茶，这是一种特别有名的绿茶。龙井茶生长在杭州周围的山丘上，以其清新、甜美的味道而闻名。虾仁便是以这种茶叶烹制而成，为这道菜带来了淡雅而芬芳的口感。

龙井虾仁已有百余年的历史，相传它曾经受到清朝乾隆皇帝的厚爱。龙井茶与虾的结合，是杭州菜肴中利用当地食材创造独特风味的典范。这道菜通过将茶的清香与虾的甜美结合，充分展示了杭州烹饪传统的精髓。

制作龙井虾仁，虽步骤简单却独具匠心。首先，要将虾清洗并剥壳。然后，将茶叶泡在热水中释放出香味。接下来，将虾仁与生姜和少许盐快速翻炒。最后，将茶叶和茶水加入锅中。虾仁吸收了茶的美味，使它们尝起来清新可口。

龙井虾仁吃起来有什么特别之处呢？龙井虾仁不仅美味，而且健康。虾仁是良好的蛋白质来源，而龙井茶富含抗氧化剂。虾仁嫩滑入味，带有淡淡的茶香。这使得这道菜既有营养又美味。让人一口就爱上了。

Dongpo Pork
东坡肉：以苏东坡命名的经典菜肴

Listening & practice   听英文原声，完成练习

▶扫码听音频◀

1. Who is Dongpo Pork named after?
   A. A famous poet and writer
   B. A famous general
   C. A famous emperor

2. What was Su Dongpo's profession (职业) besides being a talented writer?
   A. A doctor
   B. A teacher
   C. An official

3. What is the main ingredient (原料) in Dongpo Pork?
   A. Chicken breast
   B. Pork belly
   C. Beef ribs

4. How is Dongpo Pork cooked to make it tender and flavourful?
   A. It is fried quickly
   B. It is simmered slowly
   C. It is baked in an oven

5. What do people usually eat with Dongpo Pork?
   A. Noodles
   B. Bread
   C. Steamed rice

Reading 阅读下面的文章

Did you know that one of the famous dishes in traditional Chinese cuisine is named after a great writer? This dish is the renowned Dongpo Pork.

Dongpo Pork is named after Su Dongpo, a famous poet and writer from the Song Dynasty. Su Dongpo was not only talented in writing but also a master in cooking. The story goes that when Su Dongpo was an official in Xuzhou, he led the people to fight against a great flood and achieved a significant victory. To thank him, the people gave him lots of pork. Su Dongpo cooked the pork into a braised dish and shared it with the townspeople. Everyone loved it and praised his cooking skills.

Making Dongpo Pork is a unique process. First, a big piece of pork belly is cut into squares. The pork is then simmered slowly in a mix of soy sauce, sugar, ginger, and rice wine. This slow cooking makes the pork very tender and flavourful. The fat from the pork melts into the sauce, making it rich and tasty.

Dongpo Pork is not only delicious but also looks beautiful. The pork is shiny and red from the sauce, and it is served in neat squares. It's usually eaten with steamed rice, which soak up the delicious sauce.

If you like eating meat, then Dongpo Pork will surely make you happy. The pork is so tender that it almost melts in your mouth. The sweet and savory sauce adds an extra layer of flavour that makes each bite wonderful. It's a perfect dish to enjoy with your family and friends.

东坡肉：以苏东坡命名的经典菜肴

Vocabulary and phrases　词汇和短语

talented ['tæləntɪd] 形 有才能的；有天赋的

significant [sɪɡ'nɪfɪkənt] 形 重要的；有意义的

townspeople ['taʊnzpiːpl] 名 市民；镇民

simmer ['sɪmə(r)] 动 炖；煨

melt [melt] 动 融化

layer ['leɪə(r)] 名 层

achieve [ə'tʃiːv] 动 完成；达到

braised [breɪzd] 形 炖熟的；焖熟的

square [skweə(r)] 名 正方形

flavourful ['fleɪvəfʊl] 形 可口的

savory ['seɪvəri] 形 美味可口的

Practice　请选择合适的词填在下方的横线上

talented　significant　squares　flavourful　layer

1. To make Dongpo Pork, first cut the pork belly into _____.

2. My grandma's Dongpo Pork has a special _____ sauce that makes it extra delicious.

3. Dongpo Pork is a _____ dish in Chinese cuisine because it has a long history.

4. The chef who makes our Dongpo Pork is very _____.

5. When I cook Dongpo Pork, I like to add a _____ of sweet sauce on top for extra flavour.

Talking practice 情景对话模拟练习

雪莉对东坡肉很感兴趣，她请教了好朋友郑芳，让我们来一起看看这段对话吧。

Zheng Fang, I want to ask you a question. Dongpo Pork is very delicious. Is this dish related to Su Dongpo?
郑芳，我想问你一个问题。东坡肉很好吃，这道菜和苏东坡有什么关系吗？

Shirley雪莉

Zheng Fang郑芳

Yes, it is! Dongpo Pork is named after Su Dongpo, a famous poet and writer from the Song Dynasty.
是的！东坡肉是以苏东坡命名的，他是宋朝的一位著名诗人和作家。

Wow, that's interesting! Why is it named after him?
哇，真有趣！为什么以他的名字命名呢？

Shirley雪莉

Zheng Fang郑芳

Su Dongpo was also a great cook. He made this dish when he was an official in Xuzhou, and everyone loved it.
苏东坡也是一位很棒的厨师。他在徐州做官时做了这道菜，大家都很喜欢。

How is Dongpo Pork made?
东坡肉是怎么做的？

Shirley雪莉

Zheng Fang郑芳

First, the pork belly is cut into squares. Then it's simmered slowly in soy sauce, sugar, ginger, and rice wine.
首先，把五花肉切成方块。然后用酱油、糖、姜和米酒慢炖。

That sounds delicious! What does it look like?
听起来很好吃！它是什么样子的？

Shirley雪莉

东坡肉：以苏东坡命名的经典菜肴

Zheng Fang 郑芳

The pork is shiny and red from the sauce, and it's served in neat squares.
因为酱汁，所以肉看起来红亮的，被切成方块状。

Thank you, I understand!
谢谢你！我懂了！

Shirley 雪莉

Funny facts 关于东坡肉的有趣事实和短语

肉质特点：东坡肉的特点是肥而不腻，入口即化，因其长时间的炖煮，使汤汁充分渗入每一块肉中。

文学渊源：苏轼不仅是一位美食家，还是著名的文学家，他在诗词中多次提到对美食的喜爱，包括东坡肉。

slow braised – 慢炖	fatty but not greasy – 肥而不腻
yellow wine – 黄酒	melt in your mouth – 入口即化
rock sugar – 冰糖	rich flavour – 风味浓郁

Writing practice 写作小练习

根据我们这一节所学到的内容，写出以下句子的英文。

1. 东坡肉是用五花肉和特制的酱汁做成的。

   _____

2. 东坡肉的名字来源于宋朝的伟大作家苏东坡。

   _____

3. 它需要慢慢烹煮，使肉变得非常软嫩。

   _____

Reference translation 参考译文

你知道吗？中国传统菜肴中有一道菜竟是以一位伟大文人的名字命名的。这道菜便是赫赫有名的东坡肉。

东坡肉的名字来源于宋代著名的诗人和作家苏东坡。苏东坡不仅在写作上颇有才华，而且也是一个烹饪高手。传说，苏东坡在徐州当官时，曾带领全城百姓抗击洪水，取得了很大的胜利。百姓为了感谢他，纷纷送来猪肉。苏东坡把这些猪肉制作成为红烧肉，又分给了全城的百姓，大家吃完赞不绝口。

制作东坡肉是一个独特的过程。首先，将一大块五花肉切成方块。然后，将猪肉放入由酱油、糖、姜和米酒混合而成的酱汁中慢慢炖煮。这种慢炖方法使猪肉变得非常嫩滑且充满风味。猪肉中的脂肪融化到酱汁里，使其变得浓郁可口。

东坡肉不仅美味可口，而且外观也十分诱人。猪肉在酱汁的浸泡下变得红亮有光泽，被整齐地摆放在盘中。它通常与米饭一起食用，米饭可以吸收美味的酱汁。

如果你喜欢吃肉，那么东坡肉肯定会让你感到满意。猪肉嫩滑得几乎在嘴里融化。甜美和咸香的酱汁为每一口增加了一层额外的风味。与家人和朋友一起享用这道美食是再好不过的了。

Crab Meat Lion's Head
蟹粉狮子头：扬州的传统名菜

Listening & practice   听英文原声，完成练习

1. What is Crab Meat Lion's Head made from?
   A. Lion's head
   B. A mix of crab meat and pork
   C. Chicken and fish

2. Where does the name "Lion's Head" come from?
   A. The shape of the meatballs
   B. The size of the meatballs
   C. The ingredients used

3. What dynasty did Crab Meat Lion's Head originate in?
   A. Tang Dynasty
   B. Qing Dynasty
   C. Sui Dynasty

4. What makes the meatballs in Crab Meat Lion's Head tender and flavourful?
   A. Baking at high temperatures
   B. Frying in oil
   C. Simmering in a flavourful broth

5. Why is Crab Meat Lion's Head considered healthy?
   A. It is low in protein
   B. It is rich in protein from both the crab and the pork
   C. It has a lot of sugar

Reading 阅读下面的文章

Huaiyang cuisine is a famous style of Chinese cooking, and Crab Meat Lion's Head is one of its signature dishes. It originated during the Sui Dynasty, over a thousand years ago, and it is still loved by people today.

Crab Meat Lion's Head is not made from a lion's head, but rather from a mix of crab meat and pork, shaped into large meatballs. The name "Lion's Head" comes from the shape of the meatballs, which are said to look like the head of a lion. These meatballs are cooked in a tasty broth until they are tender and juicy.

Making Crab Meat Lion's Head is a fun process. First, you mix the crab meat and pork with some seasonings. Then, you shape the mixture into large meatballs. The meatballs are gently simmered in a flavourful broth made from chicken stock, ginger, and green onions. This slow cooking makes the meatballs very tender and flavourful.

Crab Meat Lion's Head is not only tasty but also healthy. It is rich in protein from both the crab and the pork. The broth adds extra flavour and makes the dish very comforting.

Eating Crab Meat Lion's Head is a delightful experience. The meatballs are soft and juicy, and the crab meat gives them a special, sweet flavour. Each bite is full of rich and savory taste, making it a perfect dish to enjoy

with your family and friends. The combination of tender meat and delicious broth makes every bite **enjoyable** and **memorable**.

Vocabulary and phrases 词汇和短语

signature dish 短 招牌菜	originate [əˈrɪdʒɪneɪt] 动 起源于
meatball [ˈmiːtbɔːl] 名 肉丸	broth [brɒθ] 名 肉汤，肉汁
seasoning [ˈsiːzənɪŋ] 名 调料	mixture [ˈmɪkstʃə(r)] 名 混合；混合物
gently [ˈdʒentli] 副 轻轻地	extra [ˈekstrə] 形 额外的
comforting [ˈkʌmfətɪŋ] 形 安慰的；令人欣慰的	enjoyable [ɪnˈdʒɔɪəbl] 形 有趣的；愉快的
memorable [ˈmemərəbl] 形 值得纪念的；难忘的	

Practice 请选择合适的词填在下方的横线上

> originated   meatball   broth   gently   enjoyable

1. Crab Meat Lion's Head is like a special kind of _____, but with crab meat inside!

2. Eating Crab Meat Lion's Head, I _____ bite into the meatball to enjoy the delicious crab meat inside.

3. The Crab Meat Lion's Head is served in a rich and flavourful _____.

4. Eating Crab Meat Lion's Head is always an _____ experience.

5. The Crab Meat Lion's Head _____ during the Sui Dynasty.

Talking practice 情景对话模拟练习

布鲁斯要去扬州旅游，出发前苏朋向他推荐了一道美食，让我们练习这段对话吧。

Su Peng, I heard you're from Yangzhou. I'm going to Yangzhou next week. Can you recommend some good food?
苏朋，听说你来自扬州，我下周要去扬州，你能帮我推荐一些美食吗？

Bruce布鲁斯

Su Peng苏朋

Sure! One of the famous dishes is Crab Meat Lion's Head.
当然可以！其中一道名菜是蟹粉狮子头。

Crab Meat Lion's Head? That sounds interesting. What is it?
蟹粉狮子头？听起来很有趣。这是什么菜？

Bruce布鲁斯

Su Peng苏朋

It's a mix of crab meat and pork shaped into large meatballs. The name comes from the shape, which looks like a lion's head.
它是由蟹肉和猪肉混合做成的大肉丸。它的名字来源于它的形状，看起来像狮子的头。

How do they make it?
他们怎么做这道菜？

Bruce布鲁斯

Su Peng苏朋

They mix crab meat and pork with seasonings, shape them into meatballs, and simmer them in a tasty broth.
他们把蟹肉和猪肉混合加调料，做成肉丸，然后在美味的汤里慢炖。

What does it taste like?
味道怎么样？

Bruce布鲁斯

蟹粉狮子头：扬州的传统名菜   035

Su Peng 苏朋

The meatballs are soft and juicy, with a special sweet flavour from the crab meat. The broth adds extra richness.
肉丸软嫩多汁，有蟹肉的特殊甜味。汤汁增加了额外的丰富口感。

That sounds great! I can't wait to try it.
听起来好棒！我迫不及待想试试了。

Bruce 布鲁斯

Funny facts 关于蟹粉狮子头的有趣事实和短语

名字由来："狮子头"因其形状大如狮头而得名，而"蟹粉"指的是其中加入的蟹肉和蟹黄，使味道更加鲜美。

烹饪方法：蟹粉狮子头通常采用炖煮的方法，将肉丸放入高汤中慢火炖煮，保持其鲜嫩多汁的口感。

文化象征：蟹粉狮子头体现了淮扬菜精致、讲究的特点，是中华美食文化的重要组成部分。

Huaiyang cuisine – 淮扬菜	crab meat – 蟹肉
pork meatball – 猪肉丸	five-spice powder – 五香粉
crab roe – 蟹黄	high protein – 高蛋白

Writing practice 写作小练习

根据我们这一节所学到的内容，写出以下句子的英文。

1. 蟹粉狮子头是由蟹肉和猪肉做成的，形状像狮子的头。

   _____

2. 把蟹肉和猪肉混合在一起，然后捏成大肉丸。

   _____

3. 蟹粉狮子头在美味的汤里煮熟。

   _____

Reference translation 参考译文

淮扬菜是中国烹饪艺术中的一颗璀璨明珠,而蟹粉狮子头则是其招牌佳肴之一。它起源于隋朝,距今已有上千年的历史,至今仍然受到人们的喜爱。

蟹粉狮子头可并不是用狮子的脑袋做的,而是用蟹肉和猪肉混合制成的大肉丸。名字中的"狮子头"来源于肉丸的形状,据说像狮子的头。这些肉丸在美味的汤汁中炖煮,直到变得嫩滑多汁。

制作蟹粉狮子头是一个有趣的过程。首先,将蟹肉和猪肉与一些调料混合在一起。然后,将混合物捏成大肉丸。肉丸在用鸡汤、姜和葱做成的美味汤汁中慢炖。这种慢炖方法使肉丸变得非常嫩滑且充满风味。

蟹粉狮子头不仅美味可口,还营养丰富。蟹肉和猪肉均为高蛋白食材,汤汁增加了额外的风味,使这道菜成为一道暖心的佳肴。

吃蟹粉狮子头是一种愉快的体验。肉丸软嫩多汁,蟹肉给它们增添了一种特别的甜味。每一口都充满了丰富而鲜美的味道,非常适合与家人和朋友一起享用。嫩滑的肉质与美味的汤汁相结合,使得每一口都令人陶醉,难以忘怀。

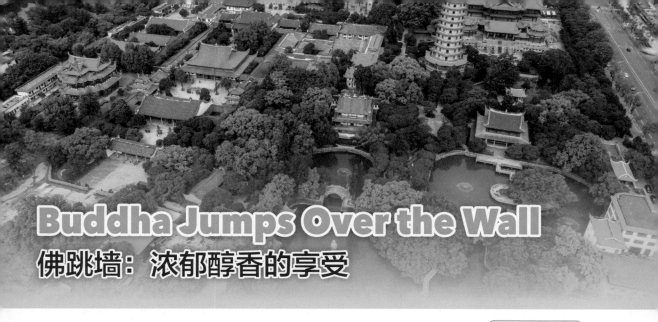

Buddha Jumps Over the Wall
佛跳墙：浓郁醇香的享受

Listening & practice 听英文原声，完成练习

▶扫码听音频◀

1. Where is Buddha Jumps Over the Wall from?
   A. Beijing
   B. Fuzhou in Fujian Province
   C. Shanghai

2. What period was Buddha Jumps Over the Wall created during?
   A. Ming Dynasty
   B. Song Dynasty
   C. Qing Dynasty

3. What did the chef add to the dish to make it even more delicious?
   A. More meat
   B. Various seafood
   C. More vegetables

4. Why is the dish called "Buddha Jumps Over the Wall"?
   A. A customer said it smelled so good that even Buddha would come to eat it
   B. It was first made in a temple
   C. It was named after a famous monk

5. What makes Buddha Jumps Over the Wall nutritious?
   A. It contains a lot of protein and vitamins from the different meats and seafood
   B. It is made with only vegetables
   C. It has a lot of sugar

Reading 阅读下面的文章

Buddha Jumps Over the Wall, also known as Fu Shou Quan, is a famous dish from Fuzhou in Fujian Province, China. It is part of the Min cuisine. There is an interesting legend about the origin of this dish.

During the Daoguang period of the Qing Dynasty, an official in Fuzhou hosted a banquet for a high-ranking official named Zhou Lian. Among the dishes served was one called "Fu Shou Quan", which was made by simmering chicken, duck, lamb shank, pig's trotters, ribs, and pigeon eggs over a slow fire. Zhou Lian was very pleased with the dish. When he returned home, he asked his chef to replicate it. The chef reduced the amount of meat and added various seafood, making the dish even more delicious and rich in flavour.

Later, Zhou Lian's chef opened his own restaurant, and this dish became the signature dish. One day, a customer exclaimed that the aroma of the dish was so enticing that even the Buddha in a nearby temple would jump over the wall to taste it. From then on, the dish was known as "Buddha Jumps Over the Wall".

Making Buddha Jumps Over the Wall is a special process. It includes a variety of ingredients such as chicken, duck, pork, lamb, and seafood like abalone, sea cucumber, and scallops. These ingredients are simmered together with Shaoxing wine and various seasonings in a clay pot. The slow cooking allows all the flavours to blend together, creating a rich and fragrant broth.

Buddha Jumps Over the Wall is not only delicious but also very nutritious. It contains a lot of protein and vitamins from the different meats and seafood. The rich broth is very comforting and flavourful. The different ingredients create a complex and delicious taste that is both savory and slightly sweet.

Vocabulary and phrases 词汇和短语

origin ['ɒrɪdʒɪn] 名 起源；出身

banquet ['bæŋkwɪt] 名 宴请；宴会

chef [ʃef] 名 厨师；主厨

exclaim [ɪk'skleɪm] 动 喝彩；大声说道

jump over 短 跳过

complex ['kɒmpleks] 名 综合体；复合体

host [həʊst] 动 做东

return [rɪ'tɜːn] 动 返回

replicate ['replɪkeɪt] 动 复制

enticing [ɪn'taɪsɪŋ] 形 引诱的；迷人的

contain [kən'teɪn] 动 容纳；包含

Practice 请选择合适的词填在下方的横线上

origin   return   chef   jumping over   contains

1. Learning about the _____ of dishes like Buddha Jumps Over the Wall helps us understand Chinese culture better.

2. After trying Buddha Jumps Over the Wall at Grandma's house, I keep asking her to _____ and make it again.

3. Even Buddha couldn't resist _____ to taste this delicious dish!

4. I love Buddha Jumps Over the Wall because it _____ all my favourite ingredients in one bowl.

5. The _____ spent hours preparing the Buddha Jumps Over the Wall.

Talking practice 情景对话模拟练习

在餐厅，一位来自国外的顾客对菜单上的佛跳墙这道菜很感兴趣，让我们跟着对话练习一下。

Customer 顾客

Hello, waiter, what is Buddha Jumps Over the Wall? Can you tell me about it?
您好，服务员，佛跳墙是什么菜？你能帮我介绍一下吗？

Waiter 服务员

Of course! Buddha Jumps Over the Wall is a famous dish from Fuzhou in Fujian Province.
当然可以！佛跳墙是福建省福州的一道名菜。

Customer 顾客

What ingredients are in it?
它有哪些配料？

Waiter 服务员

It includes chicken, duck, pork, and seafood like abalone, sea cucumber, and scallops.
它包括鸡肉、鸭肉、猪肉，还有鲍鱼、海参和扇贝等海鲜。

Customer 顾客

How is it cooked?
它是怎么做的？

Waiter 服务员

All the ingredients are simmered together with Shaoxing wine and seasonings in a clay pot.
所有配料和绍兴酒、调料一起在砂锅里慢炖。

Customer 顾客

Why is it called Buddha Jumps Over the Wall?
为什么叫佛跳墙？

Waiter 服务员

Legend says the aroma is so good that even Buddha would jump over the wall to taste it!
传说香味太好闻了，连佛祖都会跳过墙来尝一尝！

That sounds amazing! I'll definitely try it.
听起来很棒！我一定要尝尝。

Customer 顾客

Funny facts 关于佛跳墙的有趣事实和短语

食材丰富：佛跳墙的主要食材包括鲍鱼、海参、花胶、干贝、鸽蛋、鸡肉、鸭肉、猪蹄筋等。

高汤基础：高汤是佛跳墙的灵魂，通常用鸡、鸭、猪骨等熬制数小时，使汤底浓郁鲜美。

奢华象征：由于食材昂贵，制作复杂，佛跳墙常被认为是一道奢华的宴会菜肴。

Fujian cuisine – 福建菜	dried scallops – 干贝
abalone – 鲍鱼	pigeon eggs – 鸽蛋
sea cucumber – 海参	rich broth – 高汤
fish maw – 花胶	Shaoxing wine – 绍兴黄酒

Writing practice 写作小练习

根据我们这一节所学到的内容，写出以下句子的英文。

1. 佛跳墙是用很多不同的食材做成的，比如鸡肉、鸭肉和海鲜。

   _____

2. 佛跳墙的香味让佛祖都忍不住要跳墙来尝一尝。

   _____

3. 佛跳墙是中国福州的一道名菜，它真的很好吃。

   _____

Reference translation 参考译文

佛跳墙又名福寿全,是中国福建省福州市的一道特色名菜,属闽菜系。关于佛跳墙这道菜的来历,有一个有趣的传说。

清朝道光年间,福州市的一位官员请一位名叫周莲的大官吃饭,其中有一道叫作"福寿全"的菜,是用鸡、鸭、羊肘、猪蹄、排骨、鸽蛋等以慢火煨制成的。周莲对这道菜赞不绝口。回家后,他要求自己的厨师复制这道菜。厨师减少了肉的用量,并添加了各种海鲜,使这道菜更加美味,风味更加浓郁。

后来,周莲的厨师自己开了一家饭馆,这道菜成为了招牌菜。有一天,一位顾客感叹道:"这道菜的香味飘出去,寺院里的佛祖闻到了都会跳墙出来。"从此,这道菜就被叫作"佛跳墙"。

制作佛跳墙是一个特别的过程。它包含各种食材,如鸡肉、鸭肉、猪肉、羊肉以及鲍鱼、海参和扇贝等海鲜。这些食材与绍兴酒和各种调料一起在砂锅中慢炖。慢炖的过程使所有的味道融合在一起,形成了一种浓郁而芳香的汤汁。

佛跳墙不仅美味,而且非常有营养。它含有来自不同肉类和海鲜的大量蛋白质和维生素。浓郁的汤汁既舒适又美味。不同的食材创造出复杂而美味的口感,既咸鲜又略带甜味。

Boiled Chicken Slices
白切鸡：粤菜中的经典

Listening & practice 听英文原声，完成练习

▶扫码听音频◀

1. Where did Boiled Chicken Slices originate?
   A. Beijing
   B. Guangdong
   C. Sichuan

2. How is Boiled Chicken Slices cooked?
   A. It is boiled in water with ginger and green onions
   B. It is fried in oil
   C. It is baked in an oven

3. Why is the chicken cooled in ice water after cooking?
   A. To make the meat spicy
   B. To keep the meat firm and the skin smooth
   C. To add more flavour

4. What are the main ingredients in the dipping sauce for Boiled Chicken Slices?
   A. Soy sauce, ginger, garlic, and sesame oil
   B. Vinegar, sugar, and salt
   C. Peanut butter and honey

5. What makes Boiled Chicken Slices a healthy dish?
   A. It is a good source of protein and low in fat
   B. It is high in fat
   C. It is made with a lot of sugar

Reading 阅读下面的文章

Boiled Chicken Slices is a famous dish that originated in Guangdong, China. It is a classic Cantonese dish, but it has also become popular Hakka cuisine. Let's learn more about this simple but delicious dish.

Boiled Chicken Slices is made by boiling a whole chicken in water with ginger and green onions. The chicken is cooked until it is just done, making sure it stays tender and juicy. After cooking, the chicken is cooled in ice water to keep the meat firm and the skin smooth. The chicken is then cut into pieces and served with a dipping sauce.

The way Boiled Chicken Slices made is very special. By boiling the chicken slowly and carefully, the meat stays very tender. Cooling the chicken in ice water helps the skin stay smooth and adds a nice texture. This simple cooking method lets the natural flavours of the chicken shine. Unlike other dishes that use a lot of spices, Boiled Chicken Slices relies on the freshness of the chicken and a few basic ingredients to create its unique taste.

The dipping sauce for Boiled Chicken Slices is also important. It is usually made with soy sauce, ginger, garlic, and a bit of sesame oil. Sometimes, green onions and a little sugar are added to the sauce for extra flavour. The sauce adds extra flavour to the chicken and makes it even more delicious. Some people like to use a bit of chili oil to give the sauce a spicy kick.

The chicken is tender and juicy, and the skin is smooth and tasty. The dipping sauce adds a savory and slightly tangy flavour that complements the chicken perfectly. Each bite is a mix of soft, tender meat and the rich, flavourful sauce.

This dish is often served during special occasions and family gatherings. It is a symbol of purity and simplicity, reflecting the natural and wholesome qualities of the ingredients. It's a dish that is enjoyed by many people and is perfect for sharing with family and friends. Boiled Chicken Slices is not only a delicious dish but also a healthy one. Chicken is a good source of protein and is low in fat, making it a nutritious choice.

白切鸡：粤菜中的经典

Vocabulary and phrases 词汇和短语

classic ['klæsɪk] 形 经典的；传统的

simple ['sɪmpl] 形 简单的；朴素的

boil [bɔɪl] 动 煮沸

smooth [smuːð] 形 光滑的

unlike [ˌʌn'laɪk] 介 不像；与……不同

rely on 短 依靠；信赖

freshness ['freʃnəs] 名 新鲜

tangy ['tæŋi] 形 强烈的；扑鼻的

complement ['kɒmplɪment] 动 补充；补足；使完美

perfectly ['pɜːfɪktli] 副 完美地；圆满地

occasion [ə'keɪʒn] 名 场合；时机

purity ['pjʊərəti] 名 纯净；纯洁

simplicity [sɪm'plɪsəti] 名 简单；单纯

reflect [rɪ'flekt] 动 反映；反射

wholesome ['həʊlsəm] 形 有益健康的

Practice 请选择合适的词填在下方的横线上

simple   classic   smooth   Unlike   perfectly

1. Boiled Chicken Slices is a _____ Chinese dish that has been enjoyed for generations.

2. The texture of Boiled Chicken Slices is so _____, it feels like silk on my tongue.

3. Boiled Chicken Slices is a _____ but delicious dish that we often have for dinner.

4. _____ spicy hot pot, Boiled Chicken Slices is served cold and has a mild, refreshing taste.

5. The chef prepared the Boiled Chicken Slices _____, so every bite is full of flavour.

Talking practice　情景对话模拟练习

翠西和好友张琳聊起了粤菜,张琳向她推荐了白切鸡,你也跟着练习这段对话吧。

Zhang Lin, I've heard Cantonese cuisine is very famous. Are there any signature dishes?
张琳,我听说粤菜很有名,其中有什么代表菜吗?

Tracy翠西

Zhang Lin张琳

Cantonese cuisine has many delicious dishes, like roasted goose and roasted pigeon. But my top recommendation is Boiled Chicken Slices.
粤菜有很多都很好吃,比如烧鹅、烤乳鸽,不过我最推荐的就是白切鸡了。

What is Boiled Chicken Slices?
白切鸡是什么?

Tracy翠西

Zhang Lin张琳

It's made by boiling a whole chicken with ginger and green onions. The chicken is tender and juicy.
它是用姜和葱煮整只鸡。鸡肉很嫩很多汁。

How is it served?
怎么吃呢?

Tracy翠西

Zhang Lin张琳

After boiling, the chicken is cooled in ice water, then cut into pieces and served with a dipping sauce.
煮好后,鸡肉在冰水中冷却,然后切成块,配上蘸料一起吃。

What is the dipping sauce made of?
蘸料是用什么做的?

Tracy翠西

白切鸡：粤菜中的经典

Zhang Lin 张琳

It's usually made with soy sauce, ginger, garlic, and sesame oil. Sometimes, green onions and a bit of sugar are added.
通常用酱油、姜、蒜和芝麻油做的。有时候还会加上葱和一点糖。

Wow, that sounds delicious.
哇，听上去就很好吃！

Tracy 翠西

Funny facts　关于白切鸡的有趣事实和短语

名字由来：白切鸡因其制作过程中不使用酱油或其他上色调料，保持鸡肉的本色，直接切块而得名。

水煮工艺：白切鸡的制作方法是将整只鸡放入沸水中煮熟，保持鸡肉的原汁原味。

低温浸泡：鸡肉煮熟后需迅速浸泡在冷水中，使鸡皮收紧，肉质更加紧实和嫩滑。

boiled chicken – 水煮鸡	cold plunge – 冷水浸泡
tender meat – 嫩滑肉质	light seasoning – 清淡调味
ginger scallion sauce – 姜葱酱	family dish – 家常菜

Writing practice　写作小练习

根据我们这一节所学到的内容，写出以下句子的英文。

1. 白切鸡是粤菜中的经典菜肴。

   _____

2. 白切鸡不仅好吃，还很健康。

   _____

3. 我们经常在节日和家庭团聚时吃白切鸡。

   _____

Reference translation 参考译文

白切鸡起源于广东，是一道经典的粤菜，后来在客家菜中也很受欢迎。让我们一起来了解这道简单却美味的菜肴。

白切鸡是用整只鸡在姜和葱的水中煮熟而成的。煮至鸡肉恰好熟透，确保肉质保持鲜嫩多汁。煮熟后，将鸡肉放入冰水中冷却，使肉质紧实，鸡皮光滑。然后，将鸡肉切成块，并佐以特制的蘸料享用。

白切鸡的制作方式非常特别。通过精心慢煮，鸡肉保持极度的嫩滑。在冰水中冷却的过程不仅让鸡皮保持平滑，还增添了一种美妙的口感。这种简单的烹饪方法让鸡肉的天然风味得以展现。与其他使用大量香料的菜肴不同，白切鸡依赖于鸡肉的新鲜和几种基本食材来创造其独特的味道。

白切鸡的蘸料也很重要。通常是用酱油、姜、大蒜和一点芝麻油制成的。有时，还会加入葱花和一点糖来增加风味。蘸料为鸡肉增加了额外的风味，使其更加美味。有些人喜欢加入一些辣油来让蘸料带有一点辣味。

鸡肉嫩滑多汁，鸡皮光滑可口。蘸料带来咸香而略带酸爽的口感，与鸡肉完美融合。每一口都是柔软嫩滑的肉质与浓郁美味蘸料的完美交融。

这道菜常常在特殊场合和家庭聚会时供应。它象征着纯净和简单，反映了食材的天然和健康品质。这道菜深受人们的喜爱，非常适合与家人和朋友分享。白切鸡不仅美味可口，还营养丰富。鸡肉是优质蛋白质的来源，脂肪含量低，是一种营养丰富的选择。

Part 2
丰富多彩的中国地方小吃

Yangrou Paomo
羊肉泡馍：来自西北的温暖

Listening & practice 听英文原声，完成练习

▶扫码听音频◀

1. **Where is Yangrou Paomo a famous snack?**
   A. Beijing
   B. Shanghai
   C. Xi'an

2. **What does the name "Yangrou Paomo" mean?**
   A. bread soaked in lamb broth
   B. Fried lamb
   C. Spicy lamb soup

3. **What are the main ingredients in Yangrou Paomo?**
   A. Chicken and rice
   B. Lamb and flatbread
   C. Fish and noodles

4. **What is usually served on the side with Yangrou Paomo?**
   A. Pickled garlic and chili sauce
   B. Soy sauce and wasabi
   C. Ketchup and mustard

5. **What makes eating Yangrou Paomo fun and interactive?**
   A. You get to choose your own spices
   B. You get to tear the flatbread into small pieces by hand
   C. You get to cook the lamb yourself

Reading 阅读下面的文章

On a cold winter day, nothing warms you up better than a bowl of Yangrou Paomo. Yangrou Paomo is one of the most famous snacks in Xi'an, Shaanxi Province. Let's learn more about this delicious and warming snack!

Yangrou Paomo is a traditional snack from the ancient city of Xi'an, which has a rich history and was once the capital of the Tang Dynasty. Back then, Xi'an was called Chang'an. This snack is made with lamb and small pieces of flatbread soaked in a savory broth. The name "Yangrou Paomo" means "bread soaked in lamb broth", which perfectly describes how the snack is prepared and eaten.

Making Yangrou Paomo starts with a flavourful broth. The broth is made by simmering lamb with spices like star anise, cinnamon, and ginger. This slow cooking makes the broth rich and tasty. While the broth is cooking, pieces of flatbread are torn into small pieces. These bread pieces are then added into the broth, where they soak up all the delicious flavours.

The fun part about Yangrou Paomo is that you get to tear the bread yourself! When you order this snack in a restaurant, you are usually given a piece of flatbread and you tear it into small pieces by hand. This makes the meal even more special and interactive.

Yangrou Paomo tastes really good. The bread pieces become soft and flavourful after soaking in the rich broth. The tender lamb adds a delicious taste to the snack. Often, pickled garlic and chili sauce are served on the side, so you can add a bit of tangy and spicy flavour if you like.

Yangrou Paomo is not only delicious but also very satisfying. It's a perfect snack to warm you up on a cold day. People in Xi'an have enjoyed this snack for many years, and it continues to be a beloved comfort food.

Vocabulary and phrases 词汇和短语

warming ['wɔːmɪŋ] 形 暖和的；热身的	flatbread ['flætbred] 名 大饼；馍
describe [dɪ'skraɪb] 动 描述；形容	cinnamon ['sɪnəmən] 名 肉桂
soak up 短 吸收；摄取	tear [teə(r)] 动 撕
usually ['juːʒuəli] 副 通常；经常	a bit of 短 一点儿的
continue [kən'tɪnjuː] 动 继续；连续	comfort ['kʌmfət] 名 舒适；慰藉

Practice 请选择合适的词填在下方的横线上

warming   flatbread   describe   tear   a bit of

1. I can _____ Yangrou Paomo as a hearty meal.

2. When eating Yangrou Paomo, I like to _____ the flatbread into small pieces and dip them into the hot broth.

3. Yangrou Paomo comes with soft and chewy _____ that soaks up the delicious broth.

4. I like to add _____ chili oil to my Yangrou Paomo to make it a little spicy.

5. Yangrou Paomo is a _____ dish that keeps me cozy on cold days.

羊肉泡馍：来自西北的温暖 053

Talking practice 情景对话模拟练习

潘妮正在中国学习，她的母亲打电话询问了她的情况，让我们来模拟这段对话。

> Penny, did you have any interesting experiences on your trip to Xi'an?
> 潘妮，你这次去西安有什么收获吗？

Penny's Mom 潘妮的母亲

Penny 潘妮

> Yes, I visited the Terracotta Warriors and the Big Wild Goose Pagoda. And I also tried Yangrou Paomo!
> 我参观了兵马俑、大雁塔，而且我还尝试了羊肉泡馍呢！

> Yangrou Paomo? What's that?
> 羊肉泡馍？那是什么？

Penny's Mom 潘妮的母亲

Penny 潘妮

> It's a famous snack in Xi'an. It's made with lamb and pieces of flatbread soaked in a savory broth.
> 这是西安的名小吃。用羊肉和泡在美味汤里的小块馍做的。

> How do they make it?
> 他们怎么做这个小吃的？

Penny's Mom 潘妮的母亲

Penny 潘妮

> They simmer lamb with spices to make a rich broth. You tear the bread into small pieces and soak them in the broth.
> 他们用香料煮羊肉来做浓汤。你把馍撕成小块，然后泡在汤里。

> That sounds fun and delicious! Did you like it?
> 听起来很有趣也很好吃！你喜欢吗？

Penny's Mom 潘妮的母亲

Penny 潘妮

> Yes, it was really tasty and perfect for a cold day. I loved it!
> 是的，非常美味，很适合冷天吃。我很喜欢！

Funny facts 关于羊肉泡馍的有趣事实和短语

名字由来:"泡馍"指的是将馍掰成小块,泡在热腾腾的羊肉汤中,使其吸收汤汁,软嫩入味。

主要食材:羊肉泡馍的主要食材包括羊肉、馍、粉丝和各种调料,如葱、姜、蒜、香菜等。

冬季美食:羊肉泡馍因其热腾腾的特点,尤其适合在寒冷的冬季食用,具有暖身驱寒的效果。

Xi'an cuisine – 西安菜
hand-pulled bread – 手掰馍
lamb broth – 羊肉汤
garlic paste – 蒜泥
chili sauce – 辣椒酱
winter food – 冬季美食

Writing practice 写作小练习

根据我们这一节所学到的内容,写出以下句子的英文。

1. 羊肉泡馍是一道美味的中国小吃。

   _____

2. 这个小吃温暖舒适,非常适合寒冷的天气。

   _____

3. 它是由将馍浸泡在温暖的羊肉汤中制成的。

   _____

羊肉泡馍：来自西北的温暖

Reference translation 参考译文

在一个寒冷的冬日里，没有什么比一碗羊肉泡馍更能温暖你的身心了。羊肉泡馍是陕西省西安市最著名的小吃之一。让我们来更多地了解一下这道美味又暖心的小吃吧！

羊肉泡馍是古都西安的一道传统小吃，这座城市有着丰富的历史，曾经是唐朝的首都。在那个时期，西安也被叫作长安。这个小吃是用羊肉和浸泡在美味汤汁中的小块馍做成的。名字"羊肉泡馍"意为"浸泡在羊肉汤中的馍"，这完美地描述了这道菜的制作和食用方法。

制作羊肉泡馍首先需要一锅美味的汤汁。汤汁是用羊肉和八角、桂皮、姜等香料慢炖而成的。这种慢炖的方式使汤汁变得浓郁可口。在汤汁炖煮的同时，将馍撕成小块。然后，这些馍块被加入到汤汁中，充分吸收其中的美味。

羊肉泡馍的乐趣在于你可以自己撕馍！当你在餐馆点这个小吃时，通常会给你一块馍，然后你自己用手将其撕成小块。这使得这顿饭更加特别和有趣。

羊肉泡馍的味道非常好。馍块在浓郁的汤汁中浸泡后变得柔软且充满风味。嫩滑的羊肉为这个小吃增添了美味的味道。通常，泡馍会搭配腌蒜和辣椒酱一起食用，你可以根据自己的喜好增加一些酸辣的风味。

羊肉泡馍不仅美味，而且非常令人满足。它是寒冷天里温暖身体的完美小吃。西安人已经享受这个小吃很多年了，它仍然是人们喜爱的慰藉食品。

Duck Blood and Vermicelli Soup
鸭血粉丝汤：金陵的独特美味

Listening & practice 听英文原声，完成练习

▶扫码听音频◀

1. Where is Duck Blood and Vermicelli Soup a traditional dish?
   A. Beijing
   B. Shanghai
   C. Nanjing

2. What is one of the main ingredients in Duck Blood and Vermicelli Soup?
   A. Chicken blood
   B. Duck blood
   C. Pork blood

3. What type of noodles are used in Duck Blood and Vermicelli Soup?
   A. Rice noodles
   B. Wheat noodles
   C. Vermicelli noodles

4. What is added to the soup for extra flavour?
   A. Green onions and cilantro
   B. Basil and mint
   C. Parsley and thyme

5. **Why do people in Nanjing love eating Duck Blood and Vermicelli Soup?**
   A. It is very sweet
   B. It is tasty and nutritious
   C. It is cold and refreshing

Reading 阅读下面的文章

Have you ever heard of a soup made with duck blood? It might sound unusual, but in Nanjing, China, Duck Blood and Vermicelli Soup is a beloved dish! Let's explore this unique and tasty soup.

Duck Blood and Vermicelli Soup is a traditional dish from Nanjing, a city with a long history. It belongs to the Jinling cuisine, which is an important part of Nanjing's culinary tradition. This soup is made with duck blood, vermicelli noodles, tofu, and sometimes duck organs like liver and gizzard. It's one of the most famous duck-based dishes in Nanjing, renowned for its distinctive flavours.

Legend has it that during the Qing Dynasty, a man in Nanjing didn't want to waste the blood when he was slaughtering a duck, so he collected it in a small bowl. Accidentally, some vermicelli noodles fell into the bowl

and got mixed with the blood. With no other choice, he decided to cook the noodles and duck blood together. To his surprise, the result was a delicious soup with a fragrant aroma. This was how the first bowl of Duck Blood and Vermicelli Soup was created.

The soup starts with a rich broth made from duck bones. The bones are simmered with ginger, garlic, and sometimes star anise, creating a flavourful base. Duck blood is cooked and cut into cubes, which are then added to the broth along with vermicelli noodles and tofu. The soup is often garnished with green onions and cilantro for extra flavour.

Eating Duck Blood and Vermicelli Soup is an interesting experience. The vermicelli noodles are slippery and fun to eat, while the duck blood cubes are tender and smooth. The broth is rich and savory, warming you up from the inside.

In Nanjing, people love eating this soup, especially in the winter. It's not only tasty but also very nutritious. Duck blood is a good source of iron, and the soup is full of protein from the tofu and duck meat.

Trying new foods can be exciting, and Duck Blood and Vermicelli Soup is definitely worth a try. Who knows, you might discover a new favourite dish!

Vocabulary and phrases 词汇和短语

unusual [ʌn'juːʒuəl] 形 独特的；与众不同的	vermicelli [ˌvɜːmɪ'tʃeli] 名 粉丝
belong to 短 属于；为……之一员	gizzard ['gɪzəd] 名 胗
distinctive [dɪ'stɪŋktɪv] 形 独特的；与众不同的	slaughter ['slɔːtə(r)] 动 屠杀；屠宰
accidentally [ˌæksɪ'dentəli] 副 偶然地；意外地	garnish ['gɑːnɪʃ] 动 装饰
cilantro [sɪ'læntrəʊ] 名 芫荽叶；香菜	slippery ['slɪpəri] 形 滑的
source [sɔːs] 名 来源	definitely ['defɪnətli] 副 肯定地；当然地

Practice 请选择合适的词填在下方的横线上

> unusual   belongs to   accidentally   definitely   source

1. Walking by a street vendor, I _____ smelled the aroma of Duck Blood and Vermicelli Soup.

2. Trying Duck Blood and Vermicelli Soup was an _____ experience.

3. Duck Blood and Vermicelli Soup is _____ one of my favourite dishes.

4. In the world of Chinese street food, Duck Blood and Vermicelli Soup _____ the top list of must-try dishes.

5. The main _____ of flavour in Duck Blood and Vermicelli Soup comes from duck blood and broth.

Part 2 丰富多彩的中国地方小吃

Talking practice 情景对话模拟练习

鲍勃对南京的美食很感兴趣，李可给他介绍了鸭血粉丝汤，让我们来练习这段对话吧。

Li Ke, I've heard Nanjing is an ancient capital. Are there any famous snacks there?
李可，听说南京是一个古都，那里有什么比较有名的小吃吗？

Bob鲍勃

Li Ke李可
Yes, one of the famous snacks is Duck Blood and Vermicelli Soup.
是的，有一道名小吃是鸭血粉丝汤。

Duck Blood and Vermicelli Soup? That sounds unusual. What's it like?
鸭血粉丝汤？听起来很特别。它是什么样的？

Bob鲍勃

Li Ke李可
It's a soup made with duck blood, vermicelli noodles, tofu, and sometimes duck organs like liver and gizzard.
它是一种用鸭血、粉丝、豆腐，有时还有鸭肝和鸭胗做的汤。

Can you tell me how they prepare it?
你能告诉我他们是怎么做的吗？

Bob鲍勃

Li Ke李可
They start with a rich broth made from duck bones, then add cooked duck blood cubes, noodles, and tofu.
他们用鸭骨头熬制的浓汤，然后加入煮好的鸭血块、粉丝和豆腐。

What does it taste like?
味道怎么样？

Bob鲍勃

Li Ke 李可

The broth is rich and savory, the vermicelli noodles are slippery, and the duck blood cubes are tender and smooth.
汤很浓郁可口，粉丝滑溜，鸭血块嫩滑。

It sounds really unique! I'm excited to try it.
听起来真特别！我很期待尝一尝。

Bob 鲍勃

Funny facts 关于鸭血粉丝汤的有趣事实和短语

烹饪工艺：制作时，将鸭血切块，鸭肠、鸭肝、鸭胗等洗净焯水，与粉丝一起放入煮沸的鸭汤中。

口感特点：鸭血粉丝汤的口感滑嫩爽口，粉丝柔韧，鸭血细腻，汤汁鲜美，是一道非常美味的汤品。

文化象征：鸭血粉丝汤不仅是一道美食，更是南京饮食文化的重要象征之一，代表了南京人的生活方式和饮食习惯。

duck intestines – 鸭肠	green bean vermicelli – 绿豆粉丝
duck liver – 鸭肝	sweet potato vermicelli – 红薯粉丝
duck gizzard – 鸭胗	local specialty – 地方特色

Writing practice 写作小练习

根据我们这一节所学到的内容，写出以下句子的英文。

1. 鸭血粉丝汤是一种很好喝的汤。

   _____

2. 它里面有细细的粉丝和鸭血。

   _____

3. 我喜欢喝鸭血粉丝汤，因为它喝起来很暖和。

   _____

Reference translation 参考译文

你听说过用鸭血做的汤吗?这听起来可能有些奇怪,但在中国南京,鸭血粉丝汤是一道深受喜爱的美食!让我们一起来探索这道独特又美味的汤吧。

鸭血粉丝汤是南京这座拥有悠久历史的城市的传统美食。它属于金陵菜系,是南京烹饪传统的重要组成部分。这道汤是用鸭血、粉丝、豆腐,有时还有鸭的内脏如鸭肝和鸭肫制成的。它是南京最著名的鸭肉菜肴之一,以其独特的风味而闻名。

传说在清朝时期,南京的一位男子在宰杀鸭子时不想浪费鸭血,于是将其收集在一个小碗里。不小心,一些粉丝掉进了碗里,与鸭血混在了一起。无奈之下,他决定将粉丝和鸭血一起煮。令他惊讶的是,结果是一碗香气扑鼻的美味汤。这就是第一碗鸭血粉丝汤的诞生。

这道汤的汤底是用鸭骨熬制而成的浓郁高汤。鸭骨与姜、大蒜,有时还会加八角一起慢炖,形成了风味独特的底汤。鸭血煮熟后切成块,然后与粉丝和豆腐一起加入汤中。汤通常会撒上葱花和香菜来增添风味。

吃鸭血粉丝汤是一种有趣的体验。粉丝滑滑的,吃起来很有趣,而鸭血块则嫩滑顺口。汤底浓郁鲜美,从内到外温暖你的身体。

在南京,人们特别喜欢在冬天吃这道汤。它不仅美味,而且非常有营养。鸭血富含铁,汤中还含有豆腐和鸭肉提供的丰富蛋白质。

尝试新食物可以令人兴奋,而鸭血粉丝汤绝对值得一试。谁知道呢,你可能会发现一个新的最爱菜肴!

Harbin Red Sausage
哈尔滨红肠：满口都是肉香

▶扫码听音频◀

Listening & practice   听英文原声，完成练习

1. In which part of China is Harbin located?
   A. Southern China
   B. Western China
   C. Northeastern China

2. What is the main ingredient in Harbin Red Sausage?
   A. Beef
   B. Chicken
   C. Pork

3. Which country's influence can be seen in Harbin Red Sausage?
   A. France
   B. Russia
   C. Italy

4. How does Harbin Red Sausage get its red colour and smoky taste?
   A. Boiling
   B. Smoking
   C. Frying

5. How is Harbin Red Sausage often enjoyed?
   A. Grilled, fried, in soups, and in stews
   B. Raw and frozen
   C. With chocolate sauce

Reading 阅读下面的文章

In Harbin, a city known for its ice festivals and Russian style, there's a special treat that locals and visitors can't resist—Harbin Red Sausage. Let's explore what makes this sausage so unique and delicious!

Harbin Red Sausage, is a traditional sausage that originated in Harbin, a city in northeastern China. This city is famous for its Russian style, and the red sausage is a perfect example of this cultural blend. The sausage is made from pork and a variety of spices, giving it a rich and savory flavour. It's usually smoked, which adds a unique taste and makes it even more delicious.

The story of Harbin Red Sausage dates back to the early 20th century when Russian immigrants brought their sausage-making techniques to Harbin. They combined their methods with local Chinese ingredients, creating a new type of sausage that quickly became popular.

Making Harbin Red Sausage involves a few simple steps. First, the pork is finely chopped and mixed with spices like garlic, pepper, and salt. Sometimes, sugar and other seasonings are added for extra flavour. The meat mixture is then stuffed into natural casings and left to dry for a short time. After drying, the sausages are smoked over a fire, which gives them their distinctive red colour and smoky flavour.

The sausage is juicy and flavourful, with a perfect balance of spices and smokiness. It's often enjoyed as a snack, grilled or fried, and sometimes added to soups and stews for extra flavour.

In Harbin, there are many ways to enjoy this delicious sausage. Grilled over an open flame, the sausage gets a crispy exterior while remaining juicy inside, perfect for a smoky, hearty bite. Fried in a pan, it becomes golden and crunchy, ideal for a quick and tasty treat. Added to soups, the sausage imparts a rich, savory depth, making each spoonful warm and satisfying. In stews, the sausage's flavours meld with vegetables and other meats, creating a comforting and hearty meal.

Harbin Red Sausage has become a beloved part of the city's culinary heritage and a must-try for anyone visiting Harbin. It's not just a food but a cultural experience that brings warmth and flavour to cold winter days.

Vocabulary and phrases 词汇和短语

festival ['festɪvl] 名 节日；欢宴	resist [rɪ'zɪst] 动 抵抗；忍住
sausage ['sɒsɪdʒ] 名 香肠	immigrant ['ɪmɪɡrənt] 名 移民
popular ['pɒpjələ(r)] 形 受欢迎的；流行的	involve [ɪn'vɒlv] 动 包含
stuff [stʌf] 动 塞满；填满	smoky ['sməʊki] 形 冒烟的；烟熏的
smokiness [s'məʊkɪnəs] 名 烟熏味	stew [stjuː] 名 炖菜
exterior [ɪk'stɪəriə(r)] 名 外部；外皮	impart [ɪm'pɑːt] 动 赋予
spoonful ['spuːnfʊl] 名 一匙	cultural ['kʌltʃərəl] 形 文化的

Practice 请选择合适的词填在下方的横线上

> sausage   popular   smoky   stew   cultural

1. My mom often makes a _____ with Harbin Red Sausage on cold winter days.

2. Harbin Red Sausage is a special kind of _____ that comes from the city of Harbin in China.

3. I love the _____ aroma that comes from cooking Harbin Red Sausage on the grill.

4. Eating Harbin Red Sausage is a way to experience the rich _____ heritage of the Harbin region.

5. Harbin Red Sausage is a _____ choice for parties and picnics.

Part 2 丰富多彩的中国地方小吃

Talking practice 情景对话模拟练习

安娜第一次来到哈尔滨，郭玲热情地迎接了她，你也跟着练习这段对话吧。

Welcome to Harbin, a world of ice and snow.
欢迎来到哈尔滨，一个冰雪的世界。

Guo Ling郭玲

Anna安娜

Thank you, Guo Ling! I've heard Harbin has some special foods. What should I try?
谢谢你，郭玲！我听说哈尔滨有一些特别的食物。我应该尝尝什么呢？

You must try Harbin Red Sausage. It's very famous here.
你一定要尝尝哈尔滨红肠。它在这里非常有名。

Guo Ling郭玲

Anna安娜

What makes it so special?
它有什么特别之处？

It's made from pork and spices, then smoked to give it a unique flavour.
它是用猪肉和香料做的，然后熏制，给它独特的味道。

Guo Ling郭玲

Anna安娜

How do people usually eat it?
人们通常怎么吃它？

You can grill it, fry it, or add it to soups and stews. It's always delicious!
你可以烤着吃，煎着吃，或者加到汤和炖菜里。它总是很好吃！

Guo Ling郭玲

Anna安娜

That sounds amazing!
听起来很棒！

Funny facts 关于哈尔滨红肠的有趣事实和短语

历史背景：哈尔滨红肠的制作工艺最早由俄罗斯移民带入中国，结合当地口味逐渐演变为现在的哈尔滨红肠。

风味特点：哈尔滨红肠色泽红亮，外皮稍硬，切开后肉质紧实，具有浓郁的蒜香和熏香味，口感略带酸甜。

旅游特产：哈尔滨红肠是到哈尔滨旅游的游客必买的特产之一。

smoked sausage – 烟熏肠	firm texture – 紧实口感
pork sausage – 猪肉肠	traditional recipe – 传统配方
garlic flavour – 蒜香	tourist souvenir – 旅游特产

Writing practice 写作小练习

根据我们这一节所学到的内容，写出以下句子的英文。

1. 哈尔滨红肠是中国哈尔滨的一种著名食物。

   _____

2. 它是用猪肉、香料做成的，并且熏制得恰到好处。

   _____

3. 人们喜欢把哈尔滨红肠和面包一起吃。

   _____

Reference translation 参考译文

在哈尔滨这座以冰雪节和俄罗斯风情著称的城市里，有一种特别的美食让当地人和游客都欲罢不能——哈尔滨红肠。让我们来探索一下是什么让这种香肠如此独特和美味！

哈尔滨红肠，是起源于中国东北城市哈尔滨的传统香肠。这座城市以其俄罗斯风情而闻名，而红肠正是这种文化交融的完美典范。红肠是用猪肉和各种香料制成的，赋予它丰富而美味的口感。它通常是熏制的，增加了独特的风味，使其更加美味。

哈尔滨红肠的历史可以追溯到20世纪初，当时俄罗斯移民将他们的香肠制作技术带到了哈尔滨。他们将这些方法与中国本地的食材结合，创造出了一种新的香肠，迅速受到欢迎。

制作哈尔滨红肠包括几个简单的步骤。首先，将猪肉精细切碎并与大蒜、胡椒和盐等香料混合。有时还会加入糖和其他调味料以增加风味。然后将肉混合物装入天然肠衣中，并让其短时间干燥。风干后，香肠会在火上熏制，赋予其独特的红色外观和烟熏风味。

这种香肠多汁且风味浓郁，香料和烟熏味达到了完美的平衡。它经常被当作小吃享用，可以烤制或煎制，有时也会加入汤和炖菜中增加风味。

在哈尔滨，有很多种享受这种美味香肠的方法。用明火烤制时，外皮变得酥脆，而内部依然多汁，适合喜欢烟熏味的饱满一口。在平底锅中煎炸，香肠变得金黄酥脆，是快速享用美味佳肴的理想选择。加入汤中，红肠为汤增添了浓郁、美味的口感，使每一勺汤都温暖而满足。炖菜中，香肠的风味与蔬菜和其他肉类融合，创造出一种令人安慰且丰富的佳肴。

哈尔滨红肠已成为这座城市烹饪遗产中备受喜爱的一部分，也是任何到访哈尔滨的人必尝的美食之一。它不仅是一种食物，更是一种文化体验，为寒冷的冬日带来温暖和美味。

Shanghai Soup Dumpling
上海小笼包：充满鲜美的汤汁

Listening & practice 听英文原声，完成练习

▶扫码听音频◀

1. What is another name for Shanghai Soup Dumplings?
   A. Dumplings
   B. Xiaolongbao
   C. Spring Rolls

2. What makes Shanghai Soup Dumplings unique?
   A. They are filled with hot, tasty soup
   B. They are fried
   C. They are made of rice

3. Where did the story of Xiaolongbao begin?
   A. Beijing
   B. Nanxiang
   C. Guangzhou

4. What is often used to make the broth inside Xiaolongbao?
   A. Chicken skin and bones
   B. Pork skin and bones
   C. Fish skin and bones

5. How should you eat Xiaolongbao to avoid burning yourself?
   A. Take a big bite right away
   B. Eat them with a fork and knife
   C. Place it on a spoon, take a small bite to let the steam out, sip the soup, and then eat the rest

Reading 阅读下面的文章

In the busy city of Shanghai, there is a special food that everyone loves—Shanghai Soup Dumplings, also known as Xiaolongbao. Let's find out why these little dumplings are so amazing!

Shanghai Soup Dumplings are famous in Shanghai. They are unique because they are filled with hot, tasty soup! These dumplings are made from a thin dough wrapper filled with minced pork, and sometimes crab meat, mixed with a rich broth. When you bite into a Xiaolongbao, you get a burst of delicious soup along with the tender meat filling.

The story of Xiaolongbao begins in Nanxiang, a town near Shanghai. More than a hundred years ago, a clever chef created these dumplings by adding jelly-like broth to the filling. When steamed, the broth melts, creating a yummy soup inside the dumpling. This idea quickly became popular, and Xiaolongbao soon spread throughout Shanghai and beyond.

Making Shanghai Soup Dumplings is an art. First, the dough is rolled into thin circles. The filling is prepared by mixing minced pork or crab with a jelly-like broth made from pork skin and bones. This filling is then carefully wrapped in the dough, and each dumpling is folded with many small pleats to hold the soup inside. The dumplings are steamed in bamboo baskets until they are perfectly cooked.

Eating Xiaolongbao is a fun experience. You have to be careful because the soup inside is very hot. The best way to eat them is to place the dumpling on a spoon, take a small bite to let the steam out, sip the soup, and then enjoy the rest of the dumpling. Often, Xiaolongbao is served with a dipping sauce made of vinegar and ginger, which makes them even more tasty.

Shanghai Soup Dumplings are not only delicious but also a fun food to enjoy. The combination of the soft wrapper, tasty filling, and hot soup makes every bite a wonderful surprise. It's no wonder these dumplings are a favourite among locals and visitors alike.

Vocabulary and phrases 词汇和短语

find out 短 发现；找出来	dough [dəʊ] 名 生面团
minced [mɪnst] 形 切碎的；切成末的	sometimes ['sʌmtaɪmz] 副 有时
bite [baɪt] 动 咬	melt [melt] 动 融化
throughout [θruːˈaʊt] 介 遍及；贯穿	circle ['sɜːkl] 名 圆圈；圈子
pleat [pliːt] 名 褶；褶状物	wonderful ['wʌndəfl] 形 精彩的；极好的

Practice 请选择合适的词填在下方的横线上

find out   dough   melt   Throughout   wonderful

1. I can't wait to _____ what's inside a Shanghai Soup Dumpling when I bite into it!

2. The broth inside Shanghai Soup Dumplings is so flavourful that it seems to _____ on my tongue.

3. Every time I eat Shanghai Soup Dumplings, I feel so _____ and happy.

4. Making Shanghai Soup Dumplings requires carefully wrapping the filling in thin layers of _____.

5. _____ the year, people in Shanghai enjoy eating Soup Dumplings as a special treat.

Part 2 丰富多彩的中国地方小吃

Talking practice 情景对话模拟练习

文森特想去上海旅游，陈东帮他推荐了一些好玩的地方和美食，让我们跟着这段对话练习一下。

Chen Dong, what are some fun places to visit in Shanghai?
陈东，上海有哪些好玩的地方？

Vincent 文森特

Chen Dong 陈东

There are many fun places in Shanghai. You can visit the Bund or go to Disneyland.
上海好玩的地方可多了，你可以去外滩看一看，或是去迪士尼乐园玩一玩。

That sounds great! What about food? Any recommendations?
听起来很棒！那食物呢？有什么推荐的吗？

Vincent 文森特

Chen Dong 陈东

You must try Shanghai Soup Dumplings, also known as Xiaolongbao.
你一定要尝尝上海小笼包。

Why are they so popular?
为什么它们这么受欢迎？

Vincent 文森特

Chen Dong 陈东

They are filled with hot, tasty soup and minced pork. When you bite into them, you get a burst of delicious soup.
它们里面有热腾腾的美味汤汁和猪肉馅。你咬一口，就能尝到美味的汤。

What's the best way to eat them?
吃它们的最佳方法是什么？

Vincent 文森特

上海小笼包：充满鲜美的汤汁　073

Place them on a spoon, take a small bite to let the steam out, sip the soup, and then eat the rest.
把小笼包放在勺子上，咬一个小口放出蒸汽，喝汤，然后再吃剩下的部分。

Chen Dong 陈东

Yummy! I really want to taste them soon.
真好吃！我真的很想快点尝尝。

Vincent 文森特

Funny facts　关于上海小笼包的有趣事实和短语

名字由来：小笼包因其采用小笼屉蒸制而得名，"小笼"指的是小竹笼屉，"包"指的是包子。

汤汁丰富：上海小笼包以其丰富的汤汁而著称，包子皮薄馅多，一口咬下去汤汁四溢，因此又被称为"汤包"。

吃法讲究：吃小笼包有特定的方式，通常先咬一个小口，先吸掉汤汁，然后再吃包子，这样避免烫伤。

pork filling – 猪肉馅	dough preparation – 和面
thin dough wrapper – 薄面皮	filling preparation – 调馅
steaming basket – 蒸笼	street food – 街头小吃
savory soup – 鲜美汤汁	

Writing practice　写作小练习

根据我们这一节所学到的内容，写出以下句子的英文。

1. 上海小笼包是中国上海的一种受欢迎的小吃。

   _____

2. 我喜欢蘸着醋和姜汁吃小笼包。

   _____

3. 上海小笼包的馅料通常是用猪肉和蔬菜做的。

   _____

Reference translation 参考译文

在繁华的上海，有一种人人都爱的特色美食——上海小笼包。让我们来探究一下这些小笼包为何如此令人惊叹！

上海小笼包在上海很有名。它们独特的地方在于里面有热腾腾、美味的汤汁！这些小笼包由薄皮包裹着剁碎的猪肉，有时也加入蟹肉，并与浓郁的肉汤混合而成。当你咬下一口小笼包时，伴随着鲜嫩的肉馅，你会感受到一股美味的汤汁爆发出来。

小笼包的故事始于上海郊区的一个小镇——南翔。一百多年前，一位聪明的厨师通过在馅料中加入类似果冻的肉汤，创造了小笼包。当它们被蒸熟时，肉汤会融化，在小笼包内部形成美味的汤汁。这个做法很快流行起来，小笼包也迅速在上海及更远的地方传播开来。

制作上海小笼包是一门艺术。首先，将面团擀成薄圆片。馅料由剁碎的猪肉或蟹肉与用猪皮和骨头制成的类似果冻的肉汤混合而成。然后将馅料小心地包裹在面团中，每个小笼包都折叠成许多小褶皱，以锁住汤汁。小笼包被放在竹篮中蒸熟，直到完全熟透。

吃小笼包是一种有趣的体验。你要小心，因为里面的汤汁非常烫。最好的吃法是将小笼包放在汤匙上，先轻轻咬开一个小口让蒸汽散出，然后吸吮汤汁，再享用剩下的部分。小笼包通常会搭配由醋和生姜制成的蘸料，使其更加美味。

上海小笼包不仅美味，还是一种令人愉悦的美食享受。薄皮、美味的馅料和热腾腾的汤汁相结合，让每一口都充满惊喜。难怪这些小笼包深受当地人和游客的喜爱。

Guangxi Luosifen

广西螺蛳粉：有点"臭臭的"酸爽

Listening & practice   听英文原声，完成练习

▶扫码听音频◀

1. **What is Guangxi Luosifen also known as?**
   A. River snail rice noodles
   B. Chicken rice noodles
   C. Beef rice noodles

2. **What is unique about Guangxi Luosifen?**
   A. It smells a little bad but tastes very good
   B. It is very sweet and fruity
   C. It is made with chocolate

3. **How is the broth for Luosifen prepared?**
   A. By boiling chicken and vegetables
   B. By simmering river snails, pork bones, and spices for several hours
   C. By mixing soy sauce and sugar

4. **What are some common toppings for Luosifen?**
   A. Pickled bamboo shoots, peanuts, fried tofu skins, and fresh vegetables
   B. Cheese, ham, and olives
   C. Shrimp, corn, and lettuce

5. **What can you add to Luosifen if you like spicy food?**
   A. Sugar
   B. Extra chili sauce
   C. Soy sauce

Reading 阅读下面的文章

Have you ever tried a dish that's both smelly and delicious at the same time? That's the magic of Guangxi Luosifen, or river snail rice noodles! This famous dish from Liuzhou, a city in Guangxi Zhuang autonomous, has a unique taste and smell that people either love or hate.

Guangxi Luosifen is a popular street food in Liuzhou. It is made with rice noodles, a tangy and spicy broth, and various toppings like pickled bamboo shoots, peanuts, fried tofu skins, and fresh vegetables. The most interesting ingredient, however, is the river snails used to make the broth, giving it a distinct and unforgettable flavour.

The story of Luosifen began in the 1970s. It's said that a food stall owner in Liuzhou wanted to create a dish that combined the flavours of river snails and rice noodles. He experimented with different recipes and finally made a broth using river snails, which turned out to be a big hit. Over time, Luosifen gained popularity and became a beloved dish in Guangxi Zhuang autonomous and beyond.

Making Luosifen involves several steps. First, the broth is prepared by simmering river snails, pork bones, and a mix of spices for several hours. This slow cooking process creates a rich, flavourful, and slightly pungent broth. The rice noodles are then cooked and added to the broth along with the various toppings. The result is a bowl of noodles that is spicy, sour, savory, and a little bit funky.

Eating Luosifen is an adventure. The strong smell might surprise you at first, but once you take a bite, you'll experience a burst of flavours that's hard to resist. The chewy rice noodles, crunchy peanuts, and tangy pickled bamboo shoots create a delightful mix of textures. If you like spicy food, you can add extra chili sauce to make it even more exciting.

Luosifen is not only tasty but also a fun and interactive dish to enjoy. Each bowl can be customized with different toppings and levels of spiciness, making it a unique experience every time. Whether you love it or hate it, Luosifen is a must-try for anyone visiting Guangxi.

广西 螺 蛳 粉：有点"臭臭的"酸爽

Vocabulary and phrases 词汇和短语

river snail 短 田螺	various ['veəriəs] 形 各种各样的
topping ['tɒpɪŋ] 名 浇头；配菜	unforgettable [ˌʌnfə'getəbl] 形 难忘的；不会忘记的
experiment [ɪk'sperɪmənt] 动 尝试；做实验	popularity [ˌpɒpju'lærəti] 名 普及；流行
pungent ['pʌndʒənt] 形 刺鼻的；辛辣的	funky [fʌŋki] 形 新式的；稀奇古怪的
chewy ['tʃuːi] 形 耐嚼的；富有嚼劲的	customize ['kʌstəmaɪz] 动 定制

Practice 请选择合适的词填在下方的横线上

river snail   various   unforgettable   experiment   chewy

1. The noodles in Guangxi Luosifen are so _____ and delicious!

2. Guangxi Luosifen is a famous dish that uses _____ broth as its special ingredient.

3. I want to _____ with making my own Guangxi Luosifen at home!

4. Guangxi Luosifen comes with _____ toppings like peanuts, beansprouts, and crispy vegetables.

5. My first taste of Guangxi Luosifen was truly _____.

Talking practice 情景对话模拟练习

温蒂刚从广西回来，好朋友王琦和她聊起了这次旅行的收获，让我们一起来练习这段对话。

Wang Qi 王琦

Wendy, did you have any interesting experiences on your trip to Guangxi?
温蒂，你这次去广西有什么收获吗？

Wendy 温蒂

I visited the Li River and tried a special snack from Guangxi, but I forgot what it's called.
我去了漓江，并且还尝试了广西的特色小吃，但我忘了叫什么了。

Wang Qi 王琦

Is it the one that smells a bit stinky but tastes really delicious?
是不是闻着有点臭臭的，但是吃起来特别美味？

Wendy 温蒂

Yes, that's the one! What is it called?
是的，就是那个！它叫什么名字？

Wang Qi 王琦

It's called Luosifen, or river snail rice noodles. It's very popular in Guangxi.
它叫螺蛳粉，在广西非常受欢迎。

Wendy 温蒂

How is it made?
它是怎么制作出来的？

Wang Qi 王琦

The broth is made with river snails, and it has rice noodles with lots of toppings like peanuts and bamboo shoots.
汤是用田螺煮的，里面有米粉和很多配料，比如花生和竹笋。

Funny facts   关于广西螺蛳粉的有趣事实和短语

起源地：螺蛳粉起源于中国广西壮族自治区柳州市，是柳州的传统小吃之一。

酸笋特色：酸笋是螺蛳粉中重要的配料之一，具有独特的酸香味，增添了螺蛳粉的层次感。

独特口感：螺蛳粉的米粉口感顺滑，与酸笋、花生等配料混合在一起，口感丰富多样。

rice noodles – 米粉	Wood Ear (Fungus) – 木耳
pickled Bamboo Shoots – 酸笋	Chili Oil – 辣椒油
fried tofu skin – 油豆皮	

Writing practice   写作小练习

根据我们这一节所学到的内容，写出以下句子的英文。

1. 广西螺蛳粉是由米粉、螺蛳和各种香料、蔬菜做成的。
   _____

2. 它独特的味道来自香辣酸爽的汤底。
   _____

3. 如果你喜欢吃辣，不要错过广西螺蛳粉。
   _____

Reference translation   参考译文

你有没有尝试过既难闻又美味的食物？这就是广西螺蛳粉的魔力所在！这道来自广西壮族自治区柳州市的著名美食，其独特的味道和气味让人要么爱要么恨。

广西螺蛳粉是柳州的一种受欢迎的街头小吃。它是用米粉、酸辣汤底和各种配料,如酸笋、花生、炸豆皮和新鲜蔬菜制作而成的。不过,最有趣的成分是用来熬制汤底的田螺,它赋予了汤底独特而难忘的风味。

螺蛳粉的故事始于20世纪70年代。据说,柳州的一位小吃摊主想创造一道结合螺蛳和米粉味道的美食。他尝试了不同的食谱,最终用田螺熬制的汤底大获成功。随着时间的推移,螺蛳粉越来越受欢迎,成为广西壮族自治区乃至其他地方备受喜爱的美食。

制作螺蛳粉包括几个步骤。首先,将田螺、猪骨和各种香料一起熬制几个小时,制成汤底。这种慢煮过程创造出了一种浓郁、美味且略带刺鼻的汤底。然后,将米粉煮熟并加入汤底中,再添加各种配料。最终得到的是一碗集辣、酸、鲜、香于一体的米粉,还带着一丝独特的味道。

吃螺蛳粉是一种冒险体验。强烈的气味可能会让你一开始感到惊讶,但一旦你尝了一口,就会体验到难以抗拒的味道爆发。嚼劲十足的米粉、脆脆的花生和酸辣的腌竹笋创造出令人愉悦的口感。如果你喜欢辣,可以加入额外的辣椒酱来让这道美食更加刺激。

螺蛳粉不仅美味,而且是一个有趣且互动性强的美食。每碗螺蛳粉都可以根据个人喜好添加不同的配料和辣度,让每次品尝都成为一次独特的体验。无论你是喜欢它还是讨厌它,螺蛳粉都是任何到访广西的人必尝的美食。

Guangzhou Steamed Vermicelli Rolls
广州肠粉：软糯可口的早餐

Listening & practice 听英文原声，完成练习

▶扫码听音频◀

1. **What is another name for Steamed Vermicelli Rolls?**
   A. Dim Sum
   B. Changfen
   C. Spring Rolls

2. **What are Guangzhou Steamed Vermicelli Rolls made from?**
   A. Wheat flour and water
   B. Rice flour and water
   C. Corn flour and water

3. **What is a common filling for Steamed Vermicelli Rolls?**
   A. Chicken
   B. Shrimp
   C. Fish

4. **Why is the dish called "Changfen"?**
   A. Because it is rolled up like intestines
   B. Because it is made from rice
   C. Because it is steamed

5. Where can you find Steamed Vermicelli Rolls in Guangzhou?
   A. Only in hotels
   B. In bakeries
   C. In dim sum restaurants, street stalls, and specialized shops

Reading 阅读下面的文章

If you ever visit Guangzhou, there's a dish you absolutely must try—Steamed Vermicelli Rolls, also known as Changfen. This smooth and tasty treat is a favourite among locals and visitors alike.

Guangzhou Steamed Vermicelli Rolls are a classic Cantonese dish known for their soft and silky texture. They are made from a mixture of rice flour and water, which is steamed to create thin, delicate sheets. These sheets are then rolled around various fillings like shrimp, beef, pork, or vegetables and served with a light soy sauce.

Steamed Vermicelli Rolls first appeared in teahouses in Guangdong, where people have a tradition of enjoying morning tea. Over the years, Changfen has remained one of the most popular dim sum items in Cantonese morning tea. The name "Changfen" means "intestine noodle" because the rolled shape looks like intestines. But don't worry, it's just a fun name for a delicious dish!

Making Guangzhou Steamed Vermicelli Rolls is an art. First, a thin layer of rice flour batter is spread on a steaming tray. The batter is steamed until it turns into a smooth and flexible sheet. Then, the fillings are added, and the sheet is carefully rolled up. The rolls are cut into bite-sized pieces and drizzled with soy sauce before serving.

Eating Changfen is a delightful experience. The Steamed Vermicelli rolls are incredibly soft and smooth, almost melting in your mouth. The fillings add a burst of flavour, and the soy sauce enhances the overall taste. Often, Changfen is enjoyed with a cup of hot tea, making it a perfect snack or breakfast.

In Guangzhou, you can find Steamed Vermicelli Rolls in dim sum restaurants, street stalls, and even in specialized Changfen shops. They are usually served freshly steamed, ensuring that every bite is tender and flavourful. Some variations include adding a sprinkle of sesame seeds or a dash of chili sauce for extra taste.

Guangzhou Steamed Vermicelli Rolls are not only delicious but also a fun and interactive dish to enjoy. Watching the chef expertly roll and cut the rolls is fascinating, and the taste is always satisfying. It's no wonder that Changfen is a beloved dish in Guangzhou and a must-try for anyone visiting the city.

Vocabulary and phrases 词汇和短语

absolutely ['æbsəluːtli] 副 完全地；绝对地	among [ə'mʌŋ] 介 在……之中；……之一
local ['ləʊkl] 名 当地人	silky ['sɪlki] 形 丝滑的
sheet [ʃiːt] 名 薄片	intestine [ɪn'testɪn] 名 肠
tray [treɪ] 名 托盘；盘	flexible ['fleksəbl] 形 柔韧的；有弹性的
roll up 短 卷起	drizzle ['drɪzl] 动 细雨般地洒下
burst [bɜːst] 名 迸发；爆发	stall [stɔːl] 名 摊位；货摊
variation [ˌveəri'eɪʃn] 名 变种；变化	sprinkle ['sprɪŋkl] 名 少量

Practice 请选择合适的词填在下方的横线上

absolutely   Among   locals   silky   rolling up

1. If you ask _____ where to find the best Guangzhou Steamed Vermicelli Roll, they'll give you great tips.

2. Guangzhou Steamed Vermicelli Roll is _____ delicious!

3. _____ all the Chinese dishes, Guangzhou Steamed Vermicelli Roll is my favourite.

4. The vendor is _____ the fresh Guangzhou Steamed Vermicelli Roll right in front of us.

5. One of the special things about Guangzhou Steamed Vermicelli Roll is its _____ texture.

Talking practice 情景对话模拟练习

莫妮卡和梓琳正在一起吃广式早茶，梓琳为莫妮卡推荐了肠粉，让我们跟着练习这段对话。

Monica, come and try this. It's one of the most famous snacks in Cantonese morning tea.
莫妮卡，你快来尝尝这个。这是广式早茶最有名的一个小吃。

Zilin梓琳

Monica莫妮卡
It tastes really good! What is it called?
它真的很好吃！它叫什么？

It's called Changfen, or Steamed Vermicelli Rolls.
它叫肠粉。

Zilin梓琳

Monica莫妮卡
How do they make it?
它是怎么制作的？

They steam a thin layer of rice flour batter, add fillings like shrimp or beef, and roll it up.
他们蒸一层薄薄的米粉糊，加上像虾或牛肉的馅料，然后卷起来。

Zilin梓琳

广州肠粉：软糯可口的早餐

Monica莫妮卡

That sounds interesting.
听起来很有意思。

Yes! The Steamed Vermicelli rolls are very smooth and soft, and the fillings add delicious flavours.
是的。肠粉非常光滑和柔软，馅料增加了美味的风味。

Zilin梓琳

Monica莫妮卡

I love it! I want to eat more.
我喜欢它！我想再吃一些。

Funny facts 关于广州肠粉的有趣事实和短语

早餐首选：在广东，肠粉是许多人的早餐首选，尤其在广州的早茶餐厅和街头小吃摊非常常见。

制作工艺：肠粉的制作需要将米浆倒在特制的平底布上，蒸熟后迅速卷起，形成薄而有弹性的肠粉皮。

丰富酱汁：广东肠粉通常搭配特制的酱汁，酱汁由生抽、老抽、糖和水调制而成，味道鲜美。

---

cantonese cuisine – 粤菜
shrimp steamed vermicelli rolls – 虾仁肠粉
beef steamed vermicelli rolls – 牛肉肠粉

char siu steamed vermicelli rolls – 叉烧肠粉
vegetarian steamed vermicelli rolls – 素肠粉
silky texture – 丝滑口感

Writing practice 写作小练习

根据我们这一节所学到的内容，写出以下句子的英文。

1. 肠粉口感柔软且有嚼劲。

2. 广州肠粉是中国广州很受欢迎的一种早餐食物。

3. 它是由薄薄的米浆薄片包裹着各种馅料做成的。

Reference translation 参考译文

如果你有机会去广州，有一道美食你绝对要尝尝——肠粉。这道顺滑可口的美食深受当地人和游客的喜爱。

广州肠粉是经典的粤菜，以其柔软光滑的口感而闻名。它是用大米粉和水混合制成的，然后蒸制成薄而精致的片状。这些片状再包裹着各种馅料，如虾仁、牛肉、猪肉或蔬菜，最后淋上淡淡的酱油即可食用。

肠粉最早出现在广东的茶楼里，那里的人有吃早茶的习惯。多年来，肠粉一直是广式早茶中最受欢迎的点心之一。"肠粉"这个名字的意思是"肠子粉"，因为卷起来的形状看起来像肠子。但不用担心，这只是对这道美食的一个有趣的称呼！

制作广州肠粉是一门艺术。首先，将一层薄薄的米粉糊铺在蒸盘上。将米糊蒸至光滑而有弹性的片状。然后，加入馅料，小心地将薄片卷起。将卷好的肠粉切成一口大小的块，淋上酱油后即可食用。

吃肠粉是一种愉快的体验。肠粉非常柔软光滑，几乎在口中融化。馅料增添了丰富的味道，而酱油则提升了整体的口感。肠粉通常搭配一杯热茶一起享用，使其成为完美的零食或早餐。

在广州，你可以在点心餐馆、街头小摊，甚至专门的肠粉店里找到肠粉。它们通常是现蒸现卖的，确保每一口都嫩滑美味。有些变化版本还会撒上芝麻或加一点辣酱以增加风味。

广州肠粉不仅美味可口，而且是一个充满乐趣和互动性的美食。看着厨师熟练地卷制和切割肠粉非常吸引人，而其味道总是令人满意。难怪肠粉是广州人喜爱的一道美食，也是任何到访广州的人必尝的美食。

Yunnan Cross-Bridge Rice Noodles

云南过桥米线：独特的米线体验

Listening & practice 听英文原声，完成练习

▶扫码听音频◀

1. Where did Cross-Bridge Rice Noodles originate?
   A. Beijing
   B. Yunnan
   C. Shanghai

2. According to the legend, why is the dish called "Cross-Bridge Rice Noodles"?
   A. Because the scholar's wife crossed a bridge to bring him food
   B. Because the noodles are very long
   C. Because the noodles are shaped like a bridge

3. What keeps the broth hot in Cross-Bridge Rice Noodles?
   A. A thick layer of chicken fat
   B. A special pot
   C. Adding hot water

4. How are the ingredients cooked in Cross-Bridge Rice Noodles?
   A. They are cooked in a pan
   B. They are added to the hot broth and cook instantly
   C. They are boiled separately

5. Where can you find Cross-Bridge Rice Noodles in Yunnan?
   A. Only in large hotels
   B. In many restaurants and food stalls
   C. Only in special markets

Reading 阅读下面的文章

Have you ever heard of rice noodles that can cross a bridge? Do rice noodles have legs? Let's find out about Yunnan's most famous snack—Cross-Bridge Rice Noodles!

Yunnan Cross-Bridge Rice Noodles have been around for over a hundred years and have a **touching** story behind them. It is said that during the Qing Dynasty, there was a **scholar** who studied on an **island** in the middle of a lake outside the city. His wife would walk across a bridge every day to bring him lunch, but by the time she arrived, the food was often cold.

One day, she discovered that the thick layer of chicken fat on top of the chicken soup acted like a **lid**, keeping the soup warm. She realized that if she added the ingredients and rice noodles to the hot soup just before eating, the meal would taste even better. This clever **method** quickly caught on and became very popular. To honour this wise and loving wife, the dish was named "Cross-Bridge Rice Noodles".

Making Cross-Bridge Rice Noodles is quite an experience. First, a rich and hot broth is prepared using chicken, pork bones, and various spices. The broth is kept very hot and served in a large bowl. On a **separate** plate, there are **thinly** sliced raw meats, **quail** eggs, vegetables, tofu, and, of course, the rice noodles. When ready to eat, you add the raw ingredients to the steaming broth, where they cook **instantly**. The hot broth keeps the food warm and cooks the ingredients quickly, making each bite fresh and flavourful.

Eating Cross-Bridge Rice Noodles is a fun and interactive experience. You can customize your bowl with your favourite ingredients, adding as much or as little as you like. The combination of the hot, savory broth, tender noodles, and fresh ingredients creates a **delightful** and satisfying meal.

云南过桥米线：独特的米线体验

In Yunnan, you can find Cross-Bridge Rice Noodles in many restaurants and food stalls. Each place might have its own variation, but the essence of the dish **remains** the same—fresh ingredients cooked in hot, flavourful broth. Some variations even include unique local ingredients like wild mushrooms or herbs.

Cross-Bridge Rice Noodles is not only a delicious dish but also a part of Yunnan's cultural **heritage**. The story, the preparation, and the flavours all come together to make this dish special. If you ever visit Yunnan, trying Cross-Bridge Rice Noodles is a must!

Vocabulary and phrases 词汇和短语

touching ['tʌtʃɪŋ] 形 动人的；感人的	scholar ['skɒlə(r)] 名 学者
island ['aɪlənd] 名 岛	lid [lɪd] 名 盖子
method ['meθəd] 名 方式；方法	separate ['sepərət] 形 分开的；单独的
thinly ['θɪnli] 副 薄薄地；细细地	quail [kweɪl] 名 鹌鹑
instantly ['ɪnstəntli] 副 立即地；即刻地	delightful [dɪ'laɪtfl] 形 令人愉快的；可喜的
remain [rɪ'meɪn] 动 留下；保持	heritage ['herɪtɪdʒ] 名 遗产

Practice 请选择合适的词填在下方的横线上

method　　separate　　delightful　　remains　　heritage

1. Adding the _____ ingredients to the hot broth of Yunnan Cross-Bridge Rice Noodles is part of the fun of eating it.

2. The broth _____ hot until you pour it over the noodles.

3. Yunnan Cross-Bridge Rice Noodles is a delicious _____ from the Yunnan province.

4. Eating Yunnan Cross-Bridge Rice Noodles is a _____ experience every time.

5. The _____ to make Yunnan Cross-Bridge Rice Noodles is very special.

Talking practice 情景对话模拟练习

唐纳德和王凯聊起了中国小吃的有趣话题，你也跟着来练习这段对话吧。

Wang Kai, what are some Chinese snacks with interesting names?
王凯，中国有哪些小吃的名字特别有趣？

Donald 唐纳德

Wang Kai 王凯

Let me think. Oh, have you heard of Cross-Bridge Rice Noodles from Yunnan?
让我想一想。哦，你听说过云南的过桥米线吗？

What? Can rice noodles walk across a bridge by themselves?
什么？米线自己能够走过桥吗？

Donald 唐纳德

Wang Kai 王凯

No, it's just a name. There's a story behind it. A scholar's wife had to walk across a bridge to bring him food, and she made this broth to keep his food warm. So the dish got its name.
不，那只是一个名字。这背后有一个故事。一个学者的妻子每次需要走过一座桥去送食物，她做了热汤来保温他的食物。所以这道美食因此得名。

That sounds interesting. How is it made?
听起来很有意思。它是怎么制作的？

Donald 唐纳德

云南过桥米线：独特的米线体验

Wang Kai 王凯

A hot broth is served with raw ingredients like meat, vegetables, and rice noodles. You add them to the broth, and they cook instantly.

热汤配上生的食材，比如肉、蔬菜和米线。你把它们加入汤中，它们会立刻煮熟。

Wow, that must be fun to eat!
哇，那吃起来一定很有趣！

Donald 唐纳德

Wang Kai 王凯

Yes, it is. You can customize your bowl with your favourite ingredients.

是的，很有趣。你可以用你喜欢的食材来定制你的米线。

Funny facts 关于过桥米线的有趣事实和短语

地方特色：过桥米线是云南特有的美食，尤其在昆明、蒙自等地最为著名，是当地人和游客必尝的美食之一。

口感独特：过桥米线口感滑嫩，汤鲜味美，因其独特的制作方法和丰富的配料，受到食客的喜爱。

特色吃法：吃过桥米线时，先将各种生料放入滚烫的高汤中，再加入米线，最后放入配菜和调料。

rice noodles – 米线	fish slices – 鱼片
rich broth – 浓郁汤底	tofu skins – 豆腐皮
sliced chicken – 鸡片	traditional dish – 传统菜肴
sliced pork – 猪肉片	

Writing practice 写作小练习

根据我们这一节所学到的内容，写出以下句子的英文。

1. 过桥米线是将米线和生食材加到热腾腾的汤里制成的。

2. 你可以根据自己的喜好添加蔬菜、肉类和鸡蛋等配料。

3. 吃过桥米线是一种有趣的体验。

Reference translation 参考译文

你听说过能过桥的米线吗？难道米线还有腿不成？让我们一起了解一下云南最有名的小吃——过桥米线。

云南过桥米线已经有超过百年的历史，背后还藏着一个感人的故事。据说在清朝时，有一位书生在城外的湖心小岛上读书，他的妻子每天都要走过一座桥，给他送午饭，但送到时，食物常常变冷。

有一天，她发现鸡汤上覆盖着厚厚的那层鸡油有如锅盖一样，可以让汤保持温度。她意识到，如果在吃之前把配料和米线加到热汤里，饭菜的味道会更好。这个聪明的方法很快流行起来，并大受欢迎。为了纪念这位聪明而体贴的妻子，这道美食被命名为"过桥米线"。

制作过桥米线是一种特别的体验。首先，用鸡肉、猪骨和各种香料熬制出浓郁的热汤。热汤保持高温，盛在大碗里。另一个盘子上放着薄切的生肉、鹌鹑蛋、蔬菜、豆腐，当然还有米线。准备吃的时候，把生食材加到热气腾腾的汤里，它们会立刻煮熟。热汤让食物保持温暖并迅速煮熟食材，让每一口都新鲜可口。

吃过桥米线是一种有趣且互动性强的体验。你可以根据自己的喜好定制你的米线，加入你喜欢的食材，无论多少都可以。热辣可口的汤、嫩滑的米线和新鲜的食材相结合，创造出一顿令人愉悦且满足的餐食。

在云南，你可以在很多餐馆和小吃摊找到过桥米线。每个地方可能都有自己的变化版本，但这道美食的精髓保持不变——用热辣可口的汤煮新鲜食材。有些版本甚至会加入独特的当地食材，如野生蘑菇或草药。

过桥米线不仅是一道美味佳肴，而且是云南文化遗产的一部分。这个故事、制作过程和味道都使这道美食变得特别。如果你有机会到访云南，一定要尝尝这道过桥米线！

Tianjin Jianbing Guozi
天津煎饼馃子：天津的经典早餐

Listening & practice 听英文原声，完成练习

▶扫码听音频◀

1. What is Jianbing Guozi?
   A. A type of dumpling
   B. A sweet dessert
   C. A thin, savory crepe wrapped around a crispy fried dough stick

2. Where is Jianbing Guozi a popular street food?
   A. In Beijing
   B. In Tianjin
   C. In Nanjijng

3. During which dynasty did the story of Jianbing Guozi begin?
   A. The Qing Dynasty
   B. The Ming Dynasty
   C. The Han Dynasty

4. What ingredients are spread on the batter while making Jianbing Guozi?
   A. Green onions, cilantro, and sometimes sesame seeds
   B. Sugar, butter, and chocolate
   C. Cheese, ham, and olives

5. Why do people enjoy eating Jianbing Guozi?
   A. Because it is sweet
   B. Because it is fun to watch the vendors making it
   C. Because it is very cold

Reading 阅读下面的文章

Tianjin Jianbing Guozi is a beloved street food in Tianjin, known for its unique combination of a thin, savory crepe wrapped around a crispy fried dough stick. This tasty snack is perfect for breakfast.

Who invented Jianbing Guozi? There is no definite answer, but there is a widely accepted story. During the Ming Dynasty, Tianjin was an important commercial port city that attracted many merchants from all over the country, including many from Shandong. The Shandong merchants brought their pancake-making skills to Tianjin. The creative people of Tianjin combined these pancakes with the locally common breakfast food, fried dough sticks—also known as "guozi"—and thus, Jianbing Guozi was born.

Making Jianbing Guozi is a fun and skillful process. First, a thin batter made from mung bean flour and wheat flour is spread onto a hot griddle. As the batter cooks, an egg is cracked on top and spread evenly. Then, the pancake is sprinkled with green onions, cilantro, and sometimes sesame seeds for extra flavour. The most exciting part is adding Guozi, along with a savory sauce and a touch of spicy chili paste. The pancake is then folded up, creating a delicious handheld snack.

There's a reason people love eating Jianbing Guozi. The soft and chewy pancake combined with the crispy fried dough is very satisfying. The savory sauce makes each bite full of rich flavours. No wonder Jianbing Guozi is a favourite among locals and visitors!

In Tianjin, you can find Jianbing Guozi at street stalls and markets, where vendors skillfully prepare each one to order. Watching the vendors expertly spread the batter, crack the egg, and fold the pancake is a fascinating part of the experience. Each Jianbing Guozi is made fresh, ensuring that every bite is hot and flavourful.

Vocabulary and phrases 词汇和短语

crepe [kreɪp] 名 薄煎饼	invent [ɪn'vent] 动 发明;创造
answer ['ɑːnsə(r)] 名 回答;答案	commercial [kə'mɜːʃl] 形 商业的
merchant ['mɜːtʃənt] 名 商人;店主	include [ɪn'kluːd] 动 包含;包括
locally ['ləʊkəli] 副 在本地;地方性地	thus [ðʌs] 副 因此;于是
griddle ['grɪdl] 名 煎饼用浅锅	crack [kræk] 动 破碎;砸开
paste [peɪst] 名 浆糊;糊状物	handheld ['hændheld] 形 手持式的
skillfully ['skɪlfʊli] 副 巧妙地;技术好地	vendor ['vendə(r)] 名 小贩

Practice 请选择合适的词填在下方的横线上

> invented   commercial   includes   thus   skillfully

1. Tianjin Jianbing Guozi is made with fresh ingredients, _____ making it a delicious treat for everyone.

2. The chef _____ flips the crepe to ensure it's light and crispy.

3. Tianjin Jianbing Guozi was _____ a long time ago in the city of Tianjin.

4. A typical Tianjin Jianbing Guozi _____ vegetables and eggs, making it a nutritious breakfast.

5. Tianjin Jianbing Guozi, once a street food, has become a popular _____ item.

Talking practice 情景对话模拟练习

南希第一次来天津,杜娟向她推荐了天津的煎饼馃子,让我们来练习这段对话吧。

Nancy, have you tried Tianjin's Jianbing Guozi yet?
南茜,你尝试过天津的煎饼馃子了吗?

Du Juan 杜娟

Nancy 南茜

No, I haven't. What is it?
我还没有。那是什么?

It's a thin, savory crepe wrapped around a crispy fried dough stick. It's really tasty!
它是一个薄薄的咸味煎饼,里面包着一个脆脆的油条。非常好吃!

Du Juan 杜娟

Nancy 南茜

How do they make it?
它是怎么做的?

They spread a batter on a hot griddle, add an egg, sprinkle some green onions and cilantro, then add the fried dough stick with some savory sauce and chili paste.
他们把面糊摊在热锅上,加一个鸡蛋,撒上葱和香菜,然后加上油条、咸味酱和辣酱。

Du Juan 杜娟

Nancy 南茜

That sounds delicious! Where can I find it?
听起来很好吃!我在哪里能找到它?

You can find it at street stalls and markets all around Tianjin. It's a popular breakfast food.
你可以在天津的街头摊位和市场找到它。这是一种很受欢迎的早餐。

Du Juan 杜娟

Nancy 南茜

OK!
好的!

Funny facts 关于天津煎饼馃子的有趣事实和短语

名字由来:"煎饼"指的是用绿豆面粉摊成的薄饼,"馃子"指的是里面加入的炸油条,二者合在一起称为煎饼馃子。

煎饼制作:煎饼是用绿豆面粉和水调成糊,在铁板上摊成薄饼,再打入一个鸡蛋摊平,撒上葱花和香菜。

实惠便捷:煎饼馃子价格实惠,而且制作速度快,是一种便捷的街头小吃,非常适合作为早晨匆忙时的快速早餐。

Tianjin cuisine – 天津菜
savory crepe – 咸味薄饼
mung bean flour – 绿豆面粉
crispy fried dough – 脆皮
traditional snack – 传统小吃

Writing practice 写作小练习

根据我们这一节所学到的内容,写出以下句子的英文。

1. 煎饼馃子是中国的一种受欢迎的街头小吃。

2. 它是通过在热板上摊开一层薄薄的面糊制成的。

3. 煎饼馃子上通常会淋上鲜美的酱料,有时还会加一颗鸡蛋。

Reference translation 参考译文

天津煎饼馃子是天津备受欢迎的一种街头小吃，以其独特的薄而美味的煎饼包裹着香脆的炸油条的组合而闻名。这款美味的小吃非常适合作为早餐。

那么，煎饼馃子是谁发明的呢？目前还没有确切的答案，但有一个被广泛认可的说法是这样：在明朝时期，天津是一个重要的商业港口城市，吸引了大量来自全国各地的商人，其中包括不少山东商人。山东商人将家乡的煎饼制作技艺带到了天津，天津的创意人士将这些煎饼与当地常见的早餐食品——炸油条（也称为"馃子"）相结合，于是，煎饼馃子就诞生了。

制作煎饼馃子是一项有趣且需要技巧的过程。首先，将用绿豆面和小麦面制成的面糊摊在热煎饼锅上。面糊煎熟时，在上面打一个鸡蛋并均匀摊开。然后，撒上葱花、香菜，有时还会加上芝麻来增加风味。最令人兴奋的部分是加入酥脆的馃子（油条），再抹上咸味酱料和一点辣酱。最后将煎饼折叠起来，成为美味的手持小吃。

人们喜欢吃煎饼馃子是有原因的。柔软有嚼劲的煎饼和酥脆的馃子的组合，非常令人满足。而美味的酱汁，让每一口都充满了丰富的味道。难怪煎饼馃子深受当地人和游客们的喜爱。

在天津，你可以在街头小摊和市场找到煎饼馃子。摊主们会熟练地根据顾客的要求制作每一个煎饼馃子。观看摊主熟练地摊面糊、打鸡蛋和折叠煎饼是体验中非常吸引人的一部分。每个煎饼馃子都是现做的，确保每一口都是热腾腾、风味十足的。

Lanzhou Beef Noodles
兰州牛肉面：兰州的特色美食

Listening & practice 听英文原声，完成练习

▶扫码听音频◀

1. What is another name for Lanzhou Beef Noodles?
   A. Lanzhou Lamian
   B. Beijing Mian
   C. Shanghai Noodles

2. Who is said to have created Lanzhou Beef Noodles?
   A. A famous emperor
   B. A Hui chef named Ma Baozi
   C. A farmer from Gansu Province

3. What is the key ingredient for making the broth in Lanzhou Beef Noodles?
   A. Chicken bones
   B. Sheep bones
   C. Beef bones

4. What is special about the noodles in Lanzhou Beef Noodles?
   A. They are hand-pulled
   B. They are made from rice
   C. They are very short

5. What are the five essential elements of Lanzhou Beef Noodles?
   A. Clear broth, white radish slices, red chili oil, green cilantro and garlic chives, and golden noodles
   B. Clear broth, white radish slices, red chili oil, green bell peppers, and golden noodles
   C. Clear broth, white radish slices, red chili oil, green peas, and golden noodles

Reading 阅读下面的文章

Have you ever wondered what makes a bowl of noodles so special that it becomes a symbol of an entire city? In Lanzhou, there's a famous dish that captures the essence of the city in every bite—Lanzhou Beef Noodles. Let's dive into the world of this flavourful and beloved beef noodles!

Lanzhou Beef Noodles, also known as Lanzhou Lamian, is a traditional Chinese noodle soup from Lanzhou, the capital of Gansu Province. This dish is renowned for its hand-pulled noodles, rich beef broth, and vibrant toppings. It's not just a meal but a cultural experience that has been enjoyed for generations.

The story of Lanzhou Beef Noodles can be traced back to about 100 years ago. It is said that in Lanzhou, a Hui chef named Ma Baozi created this dish. He mastered the art of hand-pulling noodles and combined them with a delicious beef broth to create a meal that was both tasty and nutritious. Over time, his recipe became very famous, leading to the creation of Lanzhou Beef Noodles.

Making Lanzhou Beef Noodles is both an art and a science. The process starts with hand-pulling the noodles, a skill that requires precision and practice. The dough is stretched and pulled into long, thin noodles, each with a perfect chewy texture. The broth is made by simmering beef bones with a mix of spices, including star anise, cinnamon, and dried orange peel, creating a rich and aromatic soup.

What sets Lanzhou Beef Noodles apart are the five essential elements: clear broth, white radish slices, red chili oil, green cilantro and garlic chives, and the golden colour of the noodles. These elements not only create a visually appealing dish but also balance the flavours perfectly. The noodles are topped with thinly sliced beef, radish, fresh herbs, and a splash of chili oil, making each bowl a harmonious blend of tastes and textures.

Eating Lanzhou Beef Noodles is a delightful experience. The hand-pulled noodles are wonderfully chewy, the beef is tender, and the broth is savory and fragrant. The fresh herbs and chili oil add a burst of flavour that makes every bite exciting. It's no wonder this dish is a favourite among locals and visitors alike.

In Lanzhou, you can find these noodles in noodle shops and street stalls, where chefs skillfully pull the noodles to order. Watching them transform a lump of dough into long, thin noodles is a fascinating part of the experience. Each bowl is made fresh, ensuring that the noodles are perfectly cooked and the broth is piping hot.

Lanzhou Beef Noodles is not just a dish but a piece of Lanzhou's history and culture. It's a must-try for anyone visiting the city, offering a taste of local tradition and culinary artistry. So next time you're in Lanzhou, don't miss the chance to enjoy a steaming bowl of Lanzhou Beef Noodles!

Vocabulary and phrases 词汇和短语

entire [ɪnˈtaɪə(r)] 形 全部的；完整的

vibrant [ˈvaɪbrənt] 形 （色彩）鲜明的

require [rɪˈkwaɪə(r)] 动 要求；需要

stretch [stretʃ] 动 伸展

visually [ˈvɪʒuəli] 副 看得见地；视觉上地

transform [trænsˈfɔːm] 动 改变；变形

capture [ˈkæptʃə(r)] 动 捕获；占领

generation [ˌdʒenəˈreɪʃn] 名 代，一代

precision [prɪˈsɪʒn] 名 精确；精密度

element [ˈelɪmənt] 名 元素；要素

harmonious [hɑːˈməʊniəs] 形 和谐的；和睦的

offer [ˈɒfə(r)] 动 提供

Practice 请选择合适的词填在下方的横线上

entire   generation   requires   element   offers

1. Eating an _____ bowl of Lanzhou Beef Noodles is a complete sensory experience.
2. Pulling the noodles for Lanzhou Beef Noodles _____ skill and patience.
3. This traditional dish _____ a good balance of nutrition.
4. The rich and flavourful broth is an essential _____ in Lanzhou Beef Noodles.
5. Lanzhou Beef Noodles has been passed down from _____ to _____.

兰州牛肉面：兰州的特色美食

Talking practice 情景对话模拟练习

亨利和陈勇聊起了他在中国的美食观察，你也一起来模拟练习这段对话吧。

I see many cities have beef noodles restaurants. Where is this dish from?
我看到很多城市都有牛肉拉面的餐馆，这是哪里的美食？

Henry 亨利

Chen Yong 陈勇

Beef noodles are a popular snack from Lanzhou, and now they are famous all over China.
牛肉拉面是兰州的流行小吃，现在已经风靡全中国。

What's so special about Lanzhou Beef Noodles?
兰州牛肉拉面有什么特别之处？

Henry 亨利

Chen Yong 陈勇

They use hand-pulled noodles, a rich beef broth, and the Lanzhou Beef Noodles have five elements.
他们使用手工拉的面条，浓郁的牛肉汤，兰州牛肉拉面有五种元素。

What are the five elements?
五种元素是什么？

Henry 亨利

Chen Yong 陈勇

Clear broth, white radish slices, red chili oil, green cilantro and garlic chives, and the golden noodles.
清汤、白萝卜片、红油、绿香菜和蒜苗，还有金黄的面条。

It sounds delicious! I really want to try it.
听起来很好吃！我真的很想尝尝。

Henry 亨利

Chen Yong 陈勇

You should! Watching them hand-pull the noodles is also fascinating.
你应该试试！观看他们手工拉面也是很有趣的。

Funny facts 关于兰州牛肉面的有趣事实和短语

清汤见底：兰州牛肉面的汤底清澈见底，但味道鲜美浓郁，是用牛骨和牛肉熬制数小时而成。

配料讲究：传统兰州牛肉面中的配料包括白萝卜片、蒜苗和红辣油等，这些配料既增添了色彩，又提升了口感。

每日必食：在兰州，牛肉面是许多市民每日必食的早餐，有些人甚至一天吃两次，足见其受欢迎程度。

clear broth – 清汤	bone broth – 骨汤
beef slices – 牛肉片	elastic noodles – 弹性面条
white radish – 白萝卜	traditional recipe – 传统食谱

Writing practice 写作小练习

根据我们这一节所学到的内容，写出以下句子的英文。

1. 兰州牛肉面是用手工拉制的面条做成的。

   _____

2. 汤底浓郁美味，经常用牛骨熬制而成。

   _____

3. 兰州牛肉面里还会加上嫩牛肉片和新鲜蔬菜。

   _____

Reference translation 参考译文

你有没有想过，一碗面条怎么能如此特别，以至于成为一座城市的象征？在兰州，有一道著名的美食，它在每一口中都捕捉到了这座城市的精髓——这就是兰州牛肉面。让我们一起走进这道美味且备受喜爱的牛肉面的世界吧！

兰州牛肉面，也被称为兰州拉面，是甘肃省省会兰州的传统中式面汤。这道美食以手工拉面、浓郁的牛肉汤和鲜艳的配料而闻名。它不仅仅是一顿饭，更是一种代代相传的文化体验。

兰州牛肉面的历史可以追溯到大约100年前。据说是在兰州，有一位名叫马保子的回族厨师创造了这道美食。他完美地掌握了手工拉面的技巧，并将其与美味的牛肉汤相结合，制作出了这道既美味又营养的餐点。随着时间的推移，他的食谱变得非常有名，兰州牛肉面也因此诞生。

制作兰州牛肉面既是一门艺术，也是一门科学。这个过程从手工拉面开始，这是一项需要精确练习的技能。面团被拉伸并拉成又长又细的面条，每根面条都有完美的嚼劲。汤底是用牛骨和包括八角、肉桂、干橘皮在内的香料混合物慢炖而成的，创造出浓郁而芳香的汤。

兰州牛肉面的独特之处在于它的五个基本元素：清汤、白萝卜片、红辣椒油、绿香菜和蒜苗，以及金黄色的面条。这些元素不仅使这道美食在视觉上具有吸引力，而且完美地平衡了味道。面条上覆盖着薄片牛肉、萝卜、新鲜蒜苗和一抹辣椒油，使每一碗都成为一种味道和口感的和谐融合。

吃兰州牛肉面是一种愉快的体验。手工拉面非常有嚼劲，牛肉柔嫩，汤底鲜美且香气扑鼻。新鲜的蒜苗和辣椒油为每一口都增添了浓郁的风味，令人兴奋不已。难怪这道美食深受当地人和游客的喜爱。

在兰州，你可以在面馆和街头小摊找到这种面条。厨师们会熟练地根据顾客的要求拉制面条。看着他们将面团变成细长面条的过程是体验中令人着迷的一部分。每一碗都是现做的，确保面条煮得恰到好处，汤底热气腾腾。

兰州牛肉面不仅是一道美食，还是兰州历史和文化的一部分。它是任何到访该城市的人都必须尝试的美食，提供了当地传统和烹饪艺术的独特体验。所以，下次你在兰州时，不要错过享受一碗热腾腾的兰州牛肉面的机会！

Xinjiang Lamb Skewers
新疆羊肉串：香气四溢的烧烤

Listening & practice 听英文原声，完成练习

1. How are Xinjiang Lamb Skewers cooked?
    A. They are boiled
    B. They are steamed
    C. They are grilled

2. Who are known for their grilling skills in Xinjiang?
    A. Han people
    B. Uyghur people
    C. Tibetan people

3. What spices are used to marinate Xinjiang Lamb Skewers?
    A. Cumin, chili powder, and salt
    B. Garlic, onion, and pepper
    C. Soy sauce and ginger

4. Where can you find Xinjiang Lamb Skewers in Xinjiang?
    A. In bakeries
    B. At night markets, street stalls, and restaurants
    C. In coffee shops

5. What is a key feature of the flavour of Xinjiang Lamb Skewers?
    A. Sweet and sour
    B. Smoky and spicy
    C. Bland and salty

新疆羊肉串：香气四溢的烧烤

Reading 阅读下面的文章

Xinjiang Lamb Skewers, also known as "Yangrou Chuan", are not only one of the most distinctive local snacks in Xinjiang but are also famous throughout China and even around the world. These skewers are made with tender pieces of lamb, marinated in a blend of spices, and grilled over an open flame. The result is a deliciously smoky and savory treat that's perfect for any occasion.

For centuries, the Uyghur people of Xinjiang have been renowned for their grilling skills. They developed a special method of marinating and grilling lamb that gives it its unique flavour. This technique has been passed down through generations, making Xinjiang Lamb Skewers a cherished culinary tradition that has gradually become popular across the country.

Making Xinjiang Lamb Skewers starts with selecting high-quality lamb. The meat is cut into bite-sized pieces and marinated in a mixture of spices, including cumin, chili powder, and salt. The lamb is then threaded onto skewers and grilled over hot coals. As the skewers cook, the aroma of the spices and the sizzling lamb fills the air, making your mouth water in anticipation.

The lamb is juicy and tender, with a perfect balance of smoky and spicy flavours. The cumin and chili powder add a warm, earthy taste that enhances the natural richness of the lamb. Each bite is a delightful mix of textures and flavours that keeps you coming back for more.

In Xinjiang, you can find these skewers at night markets, street stalls, and restaurants. They are often cooked to order, ensuring that each skewer is fresh and hot. Watching the vendors expertly grill the lamb over an open flame is part of the fun. The skewers are usually served with a sprinkle of extra cumin and chili powder for those who like a bit more spice.

Xinjiang Lamb Skewers are more than just a tasty snack—they are a symbol of Xinjiang's rich cultural heritage. The Uyghur people's tradition of grilling lamb has become a beloved part of Chinese cuisine, enjoyed by people all over the country. Whether you're at a street market in Urumqi or a festival in Beijing, Xinjiang Lamb Skewers are a must-try.

Vocabulary and phrases 词汇和短语

skewer ['skjuːə(r)] 名 串肉扦，扦子

deliciously [dɪ'lɪʃəsli] 副 美味地；怡人地

lamb [læm] 名 羔羊肉

powder ['paʊdə(r)] 名 粉末；粉

anticipation [æn,tɪsɪ'peɪʃn] 名 预期；预料

expertly ['ekspɜːtli] 副 熟练地；巧妙地

tradition [trə'dɪʃn] 名 传统；惯例

grill [grɪl] 动（在烤架上）烤

century ['sentʃəri] 名 百年；世纪

cumin ['kʌmɪn] 名 茴香籽

sizzling ['sɪzlɪŋ] 形 发咝咝声的；热烈的

richness ['rɪtʃnəs] 名 富裕；丰富

flame [fleɪm] 名 火焰

Practice 请选择合适的词填在下方的横线上

centuries   lamb   powder   flame   tradition

1. The secret to Xinjiang Lamb Skewer's amazing taste is in the special seasoning _____!

2. Xinjiang Lamb Skewer is a beloved _____ that has been enjoyed in Xinjiang for generations.

3. The _____ used in Xinjiang Lamb Skewer is carefully selected for its flavour and tenderness.

4. Xinjiang Lamb Skewer has been enjoyed for _____.

5. Xinjiang Lamb Skewer is cooked over a hot _____.

新疆羊肉串：香气四溢的烧烤

Talking practice　情景对话模拟练习

埃里克打算去新疆旅游，在出发之前，他询问了同学郝静关于新疆的美食，让我们练习这段对话。

Hao Jing, I know Xinjiang has beautiful scenery. I want to travel there. Are there any special foods?
郝静，我知道新疆的风景很美，我想去那里旅游。那里有什么美食吗？

Eric 埃里克

Hao Jing 郝静
Yes, Xinjiang Lamb Skewers are very famous. They are delicious!
有的，新疆的羊肉串非常有名，很好吃！

What makes them so special?
它们有什么特别之处？

Eric 埃里克

Hao Jing 郝静
The lamb is marinated in spices like cumin and chili powder, then grilled over hot coals.
羊肉用孜然和辣椒粉等香料腌制，然后在炭火上烤。

That sounds tasty! Where can I try them?
听起来很好吃！我在哪里可以尝到它们？

Eric 埃里克

Hao Jing 郝静
You can find them at night markets, street stalls, and restaurants in Xinjiang.
你可以在新疆的夜市、街头摊位和餐馆找到它们。

I love spicy food. Are they very spicy?
我喜欢吃辣的。它们很辣吗？

Eric 埃里克

Hao Jing 郝静
They can be spicy, but you can also ask for less chili if you prefer.
它们可以很辣，但如果你喜欢的话，也可以要求少放辣椒。

Funny facts 关于新疆羊肉串的有趣事实和短语

提前腌制：羊肉在烤制前会进行腌制，使调料充分渗入肉中，增加其香味和嫩度。

炭火烤制：传统的新疆羊肉串使用炭火烤制，这样能够使羊肉外焦里嫩，带有独特的炭香味。

夜市文化：新疆的夜市上，羊肉串摊位随处可见，是夜市文化中不可或缺的一部分，吸引了大量食客。

cumin seasoning – 孜然调料
charcoal grill – 炭火烧烤
juicy texture – 多汁口感
barbecue stalls – 烧烤摊

Writing practice 写作小练习

根据我们这一节所学到的内容，写出以下句子的英文。

1. 新疆不仅风景很好，而且还有好吃的烤羊肉串。

2. 羊肉在高温木炭上烤制，赋予其烟熏和美味的口感。

3. 品尝新疆羊肉串是一种享受新疆美食风味的绝妙方式。

新疆羊肉串：香气四溢的烧烤

Reference translation 参考译文

新疆羊肉串，不仅是新疆最有特色的风味小吃之一，而且在整个中国乃至世界各地都非常有名。这些羊肉串是用嫩滑的羊肉块制成的，经过香料腌制后，在明火上烤制。最终呈现出一道美味、烟熏味十足且鲜美的佳肴，适合任何场合享用。

几个世纪以来，新疆的维吾尔族人一直以其烧烤技艺而闻名。他们开发出了一种特殊的腌制和烤制羊肉的方法，使其具有独特的风味。这种方法代代相传，使新疆羊肉串成为珍贵的烹饪传统并逐渐风靡全国。

制作新疆羊肉串的第一步是选择高品质的羊肉。将羊肉切成小块，用包括孜然、辣椒粉和盐在内的香料混合物腌制。然后，将羊肉穿在串上，在热炭上烤制。随着肉串的烤制，香料的香气和嗞嗞作响的羊肉味弥漫在空气中，让人垂涎欲滴。

羊肉多汁嫩滑，烟熏和辣味完美平衡。孜然和辣椒粉增添了一种温暖而浓郁的味道，增强了羊肉的天然风味。每一口都是质感和风味的完美融合，让人回味无穷。

在新疆，你可以在夜市、街头小摊和餐馆找到这些羊肉串。它们通常是现点现烤，确保每一串都是新鲜热乎的。看着摊贩们在明火上熟练地烤制羊肉串也是一种乐趣。羊肉串通常会撒上一些额外的孜然和辣椒粉，以满足喜欢更辣口味的人的需求。

新疆羊肉串不仅是一道美味的小吃，也是新疆丰富文化遗产的象征。维吾尔族人烤羊肉的传统已成为中国美食的重要组成部分，深受全国各地人们的喜爱。无论你是在乌鲁木齐的街头市场，还是在北京的节日活动上，新疆羊肉串都是必尝的美味。

Part 3
历史悠久的中国节日美食

Mooncake
月饼：中秋节的赏月佳品

Listening & practice 听英文原声，完成练习

▶扫码听音频◀

1. **When are mooncakes traditionally eaten?**
   A. During the Chinese New Year
   B. During the Dragon Boat Festival
   C. During the Mid-Autumn Festival

2. **What do mooncakes symbolize?**
   A. Wealth and power
   B. Reunion and togetherness
   C. Luck and prosperity

3. **What is a common filling in mooncakes?**
   A. Chocolate
   B. Lotus seed paste
   C. Peanut butter

4. **What element is sometimes added inside mooncakes for a savory flavour?**
   A. Salted egg yolk
   B. Cheese
   C. Garlic

5. **Why do people give mooncakes as gifts to friends and family?**
   A. To celebrate the moon's power
   B. To express love or appreciation
   C. To wish for wealth and power

Reading 阅读下面的文章

Every Mid-Autumn Festival, there's one delightful treat that you'll always find on people's tables—mooncakes. Let's learn more about this delicious and **meaningful pastry**!

Mooncakes are round pastries that come in various flavours and fillings. They are usually filled with sweet lotus seed paste, red bean paste, salted egg yolks, or five Kernel fillings, and the outside is a golden-brown **crust**. Mooncakes are not just tasty; they are also rich in cultural significance.

The origin of mooncakes has many stories, but the most widely accepted one is that mooncakes originated from the **ancient** worship of the moon. In ancient times, people believed that the moon had **mysterious** powers. During the Mid-Autumn Festival, they would hold **ceremonies** to worship the moon and pray for blessings from the moon god. Mooncakes were an important offering in these ceremonies.

Mooncakes are traditionally eaten during the Mid-Autumn Festival, which falls on the 15th day of the eighth month of the lunar calendar. This festival is a time for families to gather, **admire** the full moon, and share mooncakes. The round shape of the mooncakes **symbolizes** reunion and togetherness, making them perfect for this special occasion.

Making mooncakes is a **meticulous** process. First, the dough is prepared and rolled into thin sheets. The filling, whether it's lotus seed paste, red bean paste, or another flavour, is shaped into small balls and wrapped with the dough. The filled dough balls are then pressed into molds with **intricate** designs and baked until golden brown.

Eating mooncakes is a delightful experience. The rich and sweet filling **contrasts** beautifully with the slightly crisp crust. Some mooncakes even have a salted egg yolk inside, which adds a savory element to the sweet pastry. Each bite is a perfect blend of textures and flavours.

During the Mid-Autumn Festival, mooncakes are often given as gifts to friends and family. They come in beautifully decorated boxes, making them a thoughtful and **cherished** present. Sharing mooncakes with loved ones is a way to express love and appreciation, and to celebrate the joy of being together.

Mooncakes are more than just a pastry; they are a symbol of Chinese culture and tradition. They remind people of the importance of family, unity, and the joy of sharing. So, next time you see a mooncake, remember its rich history and the special meaning it holds.

Vocabulary and phrases 词汇和短语

meaningful ['miːnɪŋfl] 形 意味深长的；有意义的	pastry ['peɪstri] 名 油酥糕点
crust [krʌst] 名 外壳；坚硬的外壳	ancient ['eɪnʃənt] 形 古老的；古代的
mysterious [mɪ'stɪəriəs] 形 神秘的；不可思议的	ceremony ['serəməni] 名 仪式；典礼
admire [əd'maɪə(r)] 动 钦佩；赞美	symbolize ['sɪmbəlaɪz] 动 象征
meticulous [mə'tɪkjələs] 形 一丝不苟的；缜密的	intricate ['ɪntrɪkət] 形 错综复杂的
contrast ['kɒntrɑːst] 动 对比；成对照	cherished ['tʃerɪʃ] 形 珍爱的，珍藏的
appreciation [əˌpriːʃi'eɪʃn] 名 欣赏；感激	

Practice 请选择合适的词填在下方的横线上

meaningful   ancient   mysterious   admire   symbolizes

1. Mooncakes are a _____ symbol of reunion during the Mid-Autumn Festival.

2. I _____ the skill and effort that goes into making each perfect mooncake.

3. Each bite of mooncake brings us a taste of _____ China and its rich history.

4. Mooncakes are linked to a _____ legend about Chang'e, the moon goddess.

5. The round shape of mooncakes _____ reunion and togetherness.

Talking practice 情景对话模拟练习

中秋节期间，许萌和好友米兰达聊起了月饼的话题，你也来练习这段对话吧。

The Mid-Autumn Festival is here! Miranda, have you eaten any mooncakes yet?
中秋节到了！米兰达，你吃月饼了吗？

Xu Meng 许萌

Miranda 米兰达

I ate one piece, it was delicious. Is mooncake a traditional food for the Mid-Autumn Festival?
我吃了一块，很好吃。月饼是中秋节的传统食物吗？

Yes, mooncake is a special treat for the festival. It symbolize reunion and togetherness.
是的，月饼是节日的特别美食。它象征着团圆和睦。

Xu Meng 许萌

Miranda 米兰达

What are the mooncakes made of?
月饼是用什么做的？

They have a thin dough and different fillings like lotus seed paste, red bean paste, or salted egg yolk.
它们有薄薄的外皮，里面有各种馅料，比如莲蓉、红豆沙或者咸蛋黄。

Xu Meng 许萌

月饼：中秋节的赏月佳品　117

Miranda米兰达

That sounds so tasty! Do people give mooncakes as gifts?
听起来好好吃！人们会把月饼作为礼物送吗？

Yes, they are often given to friends and family in beautiful boxes.
是的，人们经常把月饼装在漂亮的盒子里送给朋友和家人。

Xu Meng许萌

Miranda米兰达

I love the idea of sharing mooncakes. It's such a nice tradition.
我喜欢分享月饼的主意。这真是一个美好的传统。

Funny facts 关于月饼的有趣事实和短语

多样口味：月饼有许多不同的口味和馅料，包括莲蓉、豆沙、五仁、蛋黄、枣泥、肉松等，满足不同口味需求。

地区差异：不同地区的月饼有不同的特色，如广式月饼以皮薄馅多著称，苏式月饼则以酥皮和多层次口感闻名。

花纹精美：月饼表面通常印有精美的花纹和图案，增加了节日的氛围。

the Mid-Autumn Festival – 中秋节	**red bean paste** – 红豆沙
lunar calendar – 农历	**five kernel fillings** – 五仁
reunion – 团圆	**salted egg yolk** – 咸蛋黄
lotus seed paste – 莲蓉	**festival food** – 节日食品

Writing practice 写作小练习

根据我们这一节所学到的内容，写出以下句子的英文。

1. 月饼有多种口味，如莲蓉和红豆沙。

2. 月饼是中秋节时人们常吃的美味圆形糕点。

3. 在中秋节期间，人们会给朋友和家人送上月饼作为礼物。

Reference translation 参考译文

每逢中秋佳节，人们的餐桌上总有一款令人愉悦的美食——月饼。接下来，让我们一起来了解这种既美味又有意义的糕点吧！

月饼是一种圆形的糕点，有各种不同的口味和馅料。它们通常以甜美的莲蓉、红豆沙、咸蛋黄或是五仁为馅，外面包裹着金黄的酥皮。月饼不仅美味，还蕴含着深厚的文化内涵。

月饼的起源有多种说法，其中最为广泛接受的是月饼起源于古代对月亮的崇拜。在古代，人们认为月亮具有神秘的力量，因此每到中秋节，便会举行祭月仪式，祈求月神保佑，而月饼则是这些仪式中不可或缺的供品。

月饼传统上是在中秋节食用的，这一天恰逢农历八月十五。中秋节是家庭团聚的时刻，人们会一起赏月、分享月饼。月饼的圆形象征着团圆和睦，使它们成为这个特殊节日的完美食物。

制作月饼是一个精细的过程。首先，准备好面团并擀成薄片。然后将莲蓉、红豆沙或其他馅料制成小球，用面片包裹。包好的面团球被放入带有精美图案的模具中，压制成型，最后烘烤至金黄色。

品尝月饼是一种美妙的体验。甜美的馅料与略微酥脆的外皮形成了鲜明对比。有些月饼里面还有咸蛋黄，增加了一种咸香的味道。每一口都是质地和风味的完美结合。

在中秋节期间，人们常常将月饼作为礼物送给亲朋好友。月饼装在精美的盒子里，成为一份用心且珍贵的礼物。与亲人分享月饼是一种表达爱和感激的方式，也是庆祝团圆和喜悦的时刻。

月饼不仅是一种糕点，更是中国文化和传统的象征。它们提醒人们家庭、团结的重要性，以及分享的快乐。因此，当你下次看到月饼时，请记得它丰富的历史背景和它所承载的特殊意义。

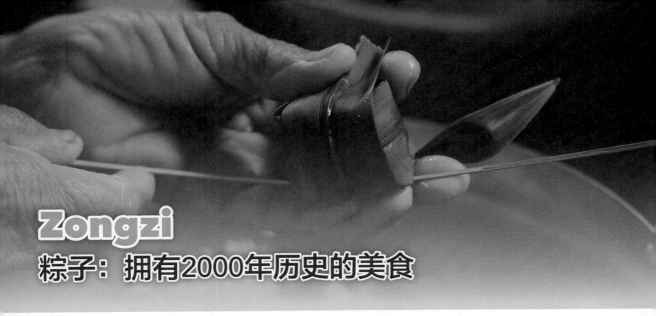

Zongzi
粽子：拥有2000年历史的美食

Listening & practice 听英文原声，完成练习

▶扫码听音频◀

1. What is Zongzi made of?
   A. Sticky rice wrapped in bamboo leaves
   B. Wheat flour
   C. Cornmeal

2. Who is the poet associated with the story of Zongzi?
   A. Li Bai
   B. Du Fu
   C. Qu Yuan

3. During which festival is Zongzi traditionally eaten?
   A. the Mid-Autumn Festival
   B. the Dragon Boat Festival
   C. the Spring Festival

4. What shape is Zongzi typically made into?
   A. Round
   B. Cube
   C. Pyramid

5. Why do people give Zongzi as gifts during the Dragon Boat Festival?
   A. To celebrate the new year
   B. To express gratitude and celebrate the spirit of the festival
   C. To wish for wealth and power

Reading 阅读下面的文章

Zongzi, also known as rice dumplings, is pyramid-shaped packet of **glutinous** rice wrapped in bamboo leaves. It comes with various fillings, such as red bean paste, jujubes, pork, and salted egg yolk. Zongzi is steamed or boiled until it becomes tender and flavourful.

The story of Zongzi can be traced back over two thousand years to the Warring States period, and it is closely associated with the poet Qu Yuan. Qu Yuan was a loyal **minister** of the Chu State. When he learned that his country was **invaded**, he was so **distraught** that he threw himself into the Miluo River. The local people, who admired Qu Yuan, threw rice into the river to feed his spirit and keep the fish away from his body. Over time, these rice offerings **evolved** into Zongzi, which are now eaten to **commemorate** Qu Yuan's **sacrifice**.

Zongzi is traditionally eaten during the Dragon Boat Festival, which falls on the fifth day of the fifth month of the lunar calendar. This festival is celebrated with dragon boat races and the making and eating of Zongzi. The **pyramid** shape of Zongzi symbolizes strength and unity, making it perfect for this festival that honours courage and **loyalty**.

How is Zongzi made? It's an interesting process. First, bamboo leaves are soaked to make them flexible. Then sticky rice is mixed with fillings and wrapped **tightly** in the leaves, forming a pyramid shape. The Zongzi is tied with string and then steamed or boiled for several hours until it is fully cooked.

Zongzi is a favourite food for all ages. The bamboo leaves infuse the glutinous rice with a subtle, earthy aroma, while the fillings add a burst of flavour. Each bite offers a combination of textures, from the soft and sticky rice to the tender meat or sweet beans inside.

During the Dragon Boat Festival, Zongzi is often given as gifts to friends and family. It comes in beautifully wrapped packages, making it a thoughtful and cherished present. Sharing Zongzi with loved ones is a way to **express gratitude** and celebrate the festival.

Zongzi is more than just a tasty treat; it is a symbol of Chinese culture and tradition. It remind people of the importance of loyalty, unity, and the joy of sharing. So, next time you enjoy zongzi, remember its rich history and the special meaning it holds.

Vocabulary and phrases 词汇和短语

glutinous ['gluːtənəs] 形 黏的；胶质的

invade [ɪn'veɪd] 动 侵略

evolve [i'vɒlv] 动 （使）逐步形成；（使）逐步演变

sacrifice ['sækrɪfaɪs] 名 牺牲

loyalty ['lɔɪəlti] 名 忠诚；忠心

express [ɪk'spres] 动 表达；表示

minister ['mɪnɪstə(r)] 名 大臣

distraught [dɪ'strɔːt] 形 心烦意乱；忧心如焚的

commemorate [kə'meməreɪt] 动 纪念

pyramid ['pɪrəmɪd] 名 棱锥体；金字塔

tightly ['taɪtli] 副 紧紧地；坚固地

gratitude ['grætɪtjuːd] 名 感激；感恩

Practice 请选择合适的词填在下方的横线上

glutinous   minister   pyramids   tightly   express

1. The Zongzi is shaped like small _____.
2. Zongzi, a traditional Chinese snack, is made with _____ rice.
3. Zongzi is wrapped _____ with leaves to keep the rice and fillings inside.

4. Qu Yuan was a famous _____ in ancient China who loved his country deeply.

5. Zongzi can _____ our love and gratitude to family during festivals.

Talking practice 情景对话模拟练习

端午节降至，布莱恩和同学徐达聊起了这个节日的话题，让我们跟着模拟这段对话吧。

I heard that the Dragon Boat Festival is coming soon. What do people in China do to celebrate it?
我听说端午节就要到了，中国人在这个节日里会有哪些庆祝活动吗？

Brian布莱恩

Xu Da徐达
We have dragon boat races and we eat Zongzi.
我们会有龙舟比赛，还会吃粽子。

What is Zongzi?
粽子是什么？

Brian布莱恩

Xu Da徐达
It is a kind of pyramid-shaped rice dumpling wrapped in bamboo leaves, with fillings like red bean paste, jujubes, pork, and salted egg yolk.
它们是用竹叶包裹的金字塔形的糯米团，有红豆沙、红枣、猪肉和咸蛋黄等馅料。

That sounds delicious! How is it made?
听起来很好吃！它是怎么做的？

Brian布莱恩

Xu Da徐达
First, bamboo leaves are soaked. Then, glutinous rice and fillings are wrapped in the leaves, tied with string, and steamed or boiled.
首先，竹叶要浸泡。然后，把糯米和馅料包裹在竹叶里，绑上绳子，蒸或煮。

粽子：拥有2000年历史的美食

Brian布莱恩

Why do people eat Zongzi during the Dragon Boat Festival?
为什么人们在端午节吃粽子？

Xu Da徐达

It's to honour the poet Qu Yuan, who drowned himself in the Miluo River. People threw rice into the river to keep the fish away from his body.
这是为了纪念诗人屈原，他投河自尽。人们把米饭扔进河里，以防鱼吃他的身体。

Brian布莱恩

Wow, Zongzi have such a meaningful history!
哇，粽子有这么有意义的历史！

Funny facts 关于粽子的有趣事实和短语

名字演变："粽子"在古代被称为"角黍"，指的是用粽叶包裹糯米制作的食品，后演变为"粽子"。

地域差异：中国各地的粽子种类丰富，北方粽子多以甜味为主，南方粽子多以咸味为主，如广东的咸肉粽、上海的碱水粽、福建的烧肉粽等。

端午习俗：除了吃粽子，端午节还有赛龙舟、挂艾草、佩香囊等传统习俗，这些习俗共同构成了端午节的丰富文化内涵。

the Dragon Boat Festival – 端午节	savory fillings – 咸馅料
glutinous rice – 糯米	red bean paste – 红豆沙
bamboo leaves – 竹叶	traditional Chinese food – 传统中国食品
sweet fillings – 甜馅料	

Writing practice 写作小练习

根据我们这一节所学到的内容，写出以下句子的英文。

1. 粽子是用竹叶包裹的糯米制成的。

2. 人们在端午节吃粽子是为了纪念诗人屈原。

3. 我妈妈常常在端午节为我煮粽子。

Reference translation 参考译文

粽子，也被称为米粽，是用竹叶包裹的糯米制成的金字塔形包裹物。粽子的馅料有很多种，比如红豆沙、枣、猪肉和咸蛋黄。粽子经过蒸煮，变得柔软且风味十足。

粽子的故事可以追溯到两千多年前的战国时期，它与诗人屈原密切相关。屈原是楚国的一位忠诚的大臣，当他得知自己的国家被侵略后，悲愤至极，投身于汨罗江。当地百姓非常敬仰屈原，于是他们将米投进江中，以祭奠他的英灵，并防止鱼儿啃食他的身体。随着时间的推移，这些米祭品演变成了今天的粽子，如今人们食用粽子，以纪念屈原的牺牲。

粽子传统上是在端午节食用的，这一天是农历五月初五。这个节日以龙舟竞赛和制作、食用粽子来庆祝。粽子的形状象征着力量与团结，非常适合这个崇尚勇气和忠诚的节日。

那么粽子是如何制作的呢？这是一个有趣的过程。首先，将竹叶浸泡使其变得柔韧。然后，将糯米和馅料混合，紧紧包裹在竹叶中，形成金字塔形。最后用绳子捆扎好，蒸或煮上几个小时，直到完全熟透。

粽子是一种老少皆宜的食物。竹叶将淡淡的清香渗入糯米中，而馅料则增添了丰富的味道。每一口都能感受到软糯的糯米和里面的肉或甜豆的完美结合。

在端午节期间，人们常常将粽子作为礼物送给亲朋好友。粽子用漂亮的包装包好，成为一份用心且珍贵的礼物。与亲人分享粽子是一种表达感激和庆祝节日的方式。

粽子不仅是一种美味的食物，更是中国文化和传统的象征。它们提醒人们忠诚、团结和分享快乐的重要性。所以，下次当你品尝粽子时，请记住它丰富的历史及其特殊的含义。

Yuanxiao
元宵：象征团圆的小吃

Listening & practice 听英文原声，完成练习

▶扫码听音频◀

1. When are yuanxiao traditionally eaten?
   A. During the Dragon Boat Festival
   B. During the Lantern Festival
   C. During the Mid-Autumn Festival

2. What are yuanxiao also known as?
   A. Tangyuan
   B. Dumplings
   C. Mooncakes

3. What is a common filling for yuanxiao?
   A. Chocolate
   B. Black sesame paste
   C. Cheese

4. What does the round shape of yuanxiao symbolize?
   A. Wealth and power
   B. Luck and prosperity
   C. Completeness and reunion

5. How are yuanxiao usually eaten?
   A. In a sweet soup
   B. With soy sauce
   C. With honey

Reading 阅读下面的文章

Have you ever celebrated the end of the Spring Festival with a special, sweet treat? In China, there's a delightful **dessert** called yuanxiao, or sweet glutinous rice balls, that is enjoyed during the Lantern Festival.

Yuanxiao, also known as tangyuan, are round, glutinous rice balls that come in various flavours and fillings. This kind of sweet treat is usually filled with ingredients like black sesame paste, red bean paste, peanut butter, or even fruit preserves. This sweet treat is often served in a light, sweet broth or soup.

The history of the Lantern Festival can be traced back over a thousand years to the Han Dynasty. It is said that the festival was **established** to promote unity and togetherness. Starting from the Song Dynasty, yuanxiao **gradually** became the symbolic food of this festival because its round shape **represents** completeness and **reunion**.

The Lantern Festival falls on the 15th day of the first month of the lunar calendar. On this night, families gather to admire beautiful lantern displays, guess **riddles** written on lanterns, and eat yuanxiao. The round shape of the yuanxiao symbolizes the full moon, which is also celebrated during this festival. This shape also **signifies** family unity and happiness, making yuanxiao the perfect treat for this joyous occasion.

Making yuanxiao is not a complicated process. First, glutinous rice flour is mixed with water to form a smooth and **pliable** dough. Then, small pieces of dough are filled with sweet fillings and rolled into perfect round shapes. Finally, these rice balls are cooked in boiling water until they become soft and chewy.

Yuanxiao tastes really good! The glutinous rice balls have a **wonderfully** chewy texture, and the sweet fillings **provide** a burst of flavour with each bite. Whether filled with black sesame, red bean paste, or other sweet fillings, yuanxiao is loved by everyone, especially children.

During the Lantern Festival, yuanxiao is often shared with friends and family. It is sometimes given as gift, beautifully **packaged** to symbolize good wishes and harmony. Sharing yuanxiao is a way to express love and celebrate the bonds of family and friendship.

Yuanxiao is more than just a dessert; it is a symbol of Chinese culture and tradition. It reminds people of the importance of family, unity, and the joy of sharing. So, next time you enjoy yuanxiao, remember its rich history and the special meaning they hold.

Vocabulary and phrases 词汇和短语

dessert [dɪ'zɜːt] 名 甜食；甜品

gradually ['grædʒuəli] 副 逐渐地

reunion [ˌriː'juːniən] 名 团聚

signify ['sɪɡnɪfaɪ] 动 意味着；象征

wonderfully ['wʌndəfəli] 副 绝妙地；极佳地

package ['pækɪdʒ] 动 把……打包

establish [ɪ'stæblɪʃ] 动 建立；确立

represent [ˌreprɪ'zent] 动 代表；象征

riddle ['rɪdl] 名 谜；谜语

pliable ['plaɪəbl] 形 柔软的；圆滑的

provide [prə'vaɪd] 动 供给；提供

unity ['juːnəti] 名 团结；统一

Practice 请选择合适的词填在下方的横线上

dessert   reunion   riddles   wonderfully   provides

1. Eating yuanxiao during the Lantern Festival is a symbol of family _____ and happiness.

2. The sweet taste of yuanxiao _____ a wonderful symbol of happiness.

3. Yuanxiao has a soft and chewy texture, making it a delicious _____.

Part 3 历史悠久的中国节日美食

4. During the Lantern Festival, people guess _____ for fun.
5. The sweet, soft yuanxiao tastes _____ on a cold Lantern Festival night.

Talking practice　情景对话模拟练习

杰克对元宵节很感兴趣，他和好朋友李雪聊起了元宵，你也跟着模拟这段对话吧。

Hi Li Xue, I heard about the Lantern Festival. What do people do to celebrate it?
你好，李雪，我听说过元宵节。人们怎么庆祝这个节日呢？

Jack杰克

Li Xue李雪

We admire beautiful lantern displays, guess riddles, and eat yuanxiao.
我们会观赏美丽的灯笼，猜灯谜，还有吃元宵。

What is yuanxiao?
元宵是什么？

Jack杰克

Li Xue李雪

Yuanxiao is a sweet glutinous rice ball filled with ingredients like black sesame paste, red bean paste, or peanut butter.
元宵是一种甜的糯米团，里面有黑芝麻酱、红豆沙或者花生酱等馅料。

That sounds delicious! How are the sweet glutinous rice balls made?
听起来很好吃！这些甜糯米团是怎么做的？

Jack杰克

Li Xue李雪

First, we mix glutinous rice flour with water to make dough. Then we fill small pieces of dough with sweet fillings and roll them into round shapes.
首先，我们把糯米粉和水混合做成面团。然后，我们把面团分成小块，包上甜馅料，搓成圆形。

元宵：象征团圆的小吃　129

Jack 杰克

Why do people eat yuanxiao during the Lantern Festival?
为什么人们在元宵节吃元宵？

Li Xue 李雪

The round shape of yuanxiao symbolizes family unity and happiness, just like the full moon.
元宵的圆形象征着家庭团圆和幸福，就像满月一样。

Jack 杰克

Wow, yuanxiao has such a special meaning! I'd love to try it someday.
哇，元宵有这么特别的意义！我真想有一天尝尝它们。

Funny facts　关于元宵的有趣事实和短语

名字由来："元宵"得名于元宵节，即农历正月十五，这一天是春节后的第一个月圆之夜，象征团圆和美满。

制作方法：元宵的制作方法有两种，一种是"滚"元宵，把糯米粉加水制成小球，再滚入馅料；另一种是"包"元宵，将糯米粉揉成皮，包入馅料后搓圆。

red bean paste – 红豆沙	**boiled yuanxiao** – 煮元宵
peanut filling – 花生馅	**fried yuanxiao** – 炸元宵
jujube paste – 枣泥	

Writing practice　写作小练习

根据我们这一节所学到的内容，写出以下句子的英文。

1. 元宵是我们在元宵节时享用的一种甜点。

　

2. 元宵的圆形象征着幸福和团圆。

3. 我喜欢吃元宵，因为它又软又糯，非常甜。

Reference translation 参考译文

在春节结束时，你有没有吃一种特别的甜点来庆祝呢？在中国，有一种叫作元宵的美味甜点，在元宵节期间享用。

元宵，也被称为汤圆，是用糯米做成的圆形甜点，有各种不同的口味和馅料。这种甜美的点心通常以黑芝麻、红豆沙、花生酱或果酱为馅，通常在淡淡的甜汤或糖水中食用。

元宵节的历史可以追溯到一千多年前的汉朝。据说元宵节的设立是为了促进团结和团圆。从宋代开始，元宵逐渐成为了这个节日的标志性食物，因为它们的圆形代表了圆满和团圆。

元宵节在农历正月十五。这天晚上，家人们聚在一起欣赏美丽的灯笼，猜灯谜，并享用元宵。元宵的圆形象征着满月，这也是这个节日庆祝的。这个形状也象征着家庭的团结和幸福，使元宵成为这个欢乐节日的完美甜点。

元宵的做法并不复杂。首先用糯米粉和水混合成光滑且柔韧的面团。然后将小块面团包上甜馅料，揉成完美的圆形。最后，把这些糯米球在沸水中煮熟，直到它们变得柔软且有嚼劲。

元宵真是美味！这些糯米球有着令人愉悦的嚼劲，而甜馅料则随着每一口都爆发出丰富的味道。无论是黑芝麻馅、红豆沙馅还是其他甜馅，元宵都深受人们的喜爱，特别是孩子们。

在元宵节期间，人们常常与朋友和家人分享元宵。它有时被作为礼物赠送，精美的包装寓意着美好的祝愿与和谐。分享元宵是表达爱意和庆祝家庭与友谊纽带的一种方式。

元宵不仅是一种甜点，更是中国文化和传统的象征。它提醒人们家庭的重要性、团结的力量以及分享的快乐。因此，下次当你品尝元宵时，请记得它丰富的历史以及它所承载的特殊意义。

Dumplings
饺子:有"年味儿"的美食

Listening & practice 听英文原声,完成练习

▶扫码听音频◀

1. What are dumplings also known as in Chinese?
   A. Tangyuan
   B. Jiaozi
   C. Baozi

2. Who is said to have invented dumplings over 1,800 years ago?
   A. Li Shizhen
   B. Sun Simiao
   C. Zhang Zhongjing

3. What does the shape of dumplings symbolize during Chinese New Year?
   A. Wealth and prosperity
   B. Health and longevity
   C. Happiness and joy

4. What is a common filling for dumplings?
   A. Ground pork with cabbage
   B. Chicken with potatoes
   C. Beef with carrots

5. How are dumplings often served?
   A. With ketchup
   B. With a dipping sauce made from soy sauce and vinegar
   C. With butter

Reading 阅读下面的文章

When Chinese New Year arrives, families across China gather to celebrate with a special meal, and one dish that always steals the **spotlight** is dumplings. Let's explore the story, tradition, and flavours of this beloved Chinese food!

Dumplings, also known as jiaozi, are a **staple** of Chinese cuisine, especially during the Spring Festival. These delicious **morsels** are made from a thin dough wrapper filled with a variety of ingredients like ground meat, vegetables, and spices. They can be boiled, steamed, or fried to **perfection**.

The origin of dumplings has many interesting stories. One such story is that dumplings were invented over 1,800 years ago by a famous Chinese doctor named Zhang Zhongjing. He made dumplings to help people **ward off** cold and illness during the harsh winter months. Eating dumplings during the Spring Festival has a long history. The shape of the dumplings, **resembling** ancient Chinese money, symbolizes wealth and **prosperity**, making them a perfect dish for welcoming the new year.

Preparing dumplings is a joyful and **communal** activity, often involving the entire family. On New Year's Eve, families gather around the table to make dumplings together, each person **contributing** to the process. The dough is rolled into thin circles, the filling is prepared, and each dumpling is carefully folded and sealed. Some families even hide a coin inside one of the dumplings—**whoever** finds it is said to have good luck in the coming year!

Dumplings have an **excellent** texture. The dough is tender and slightly chewy, while the filling is savory and flavourful. Common fillings include ground pork with cabbage, shrimp with chives, and **vegetarian** options like mustard greens. When you bite into a **freshly** cooked dumpling, the rich flavours burst in your mouth, creating a delightful taste experience.

Dumplings are often served with a dipping sauce made from soy sauce, vinegar, and sometimes a touch of chili oil or garlic. This adds an extra layer of flavour and makes each bite even more enjoyable. Whether boiled, steamed, or fried, dumplings are a **versatile** and beloved dish that brings warmth and happiness to any meal.

During the Spring Festival, dumplings are more than just food; they are a symbol of family unity and the joy of being together. Making and eating dumplings is a way to celebrate the new year, share good wishes, and create lasting memories with loved ones.

Vocabulary and phrases 词汇和短语

spotlight ['spɒtlaɪt] 名 聚光灯；公众注意的中心	staple ['steɪpl] 名 主食
morsel ['mɔːsl] 名 少量；一口	perfection [pə'fekʃn] 名 完美
ward off 短 防止；避开	resemble [rɪ'zembl] 动 与……相似；像
prosperity [prɒ'sperəti] 名 繁荣；兴旺	communal [kə'mjuːnl] 形 公有的；公社的
contribute [kən'trɪbjuːt] 动 贡献	whoever [huː'evə(r)] 代 无论是谁；不管谁
excellent ['eksələnt] 形 杰出的；优秀的	vegetarian [ˌvedʒə'teəriən] 形 素食的
freshly ['freʃli] 副 新近；刚才	versatile ['vɜːsətaɪl] 形 多功能的

Practice 请选择合适的词填在下方的横线上

ward off   contribute   whoever   excellent   Freshly

1. Dumplings are an _____ choice for any gathering because everyone loves them.

2. Eating dumplings during the Spring Festival is believed to _____ bad luck.

3. Dumplings are delicious, and _____ tries them falls in love with their taste.

4. Dumplings _____ greatly to Chinese food culture.

5. _____ made dumplings are always the best, with a delicious and juicy filling.

Talking practice 情景对话模拟练习

莎莉在中国度过了春节并在朋友家吃了饺子，让我们看看她是如何和爸爸说起这段经历的。

Sally's Dad
莎莉的爸爸

Sally, how are you doing in China?
莎莉，你在中国怎么样？

Sally莎莉

I'm doing great. It's the Spring Festival here, and I had dumplings at my friend's house.
我很好。这里正好是中国新年，我在朋友家吃了饺子。

Sally's Dad
莎莉的爸爸

I've heard of dumplings. Are they tasty?
我听说过饺子。怎么样，好吃吗？

Sally莎莉

Yes, they are delicious! The filling is very flavourful.
是的，非常好吃！饺子馅料很有味道。

Sally's Dad
莎莉的爸爸

What are they made of?
它们是用什么做的？

Sally莎莉

They are made with a thin dough wrapper and filled with meat, vegetables, and spices.
它们用薄薄的面皮包裹着肉、蔬菜和香料。

饺子：有"年味儿"的美食 135

Sally's Dad
莎莉的爸爸

It sounds like you had a wonderful New Year celebration!
听起来你过了一个很棒的新年！

Sally莎莉

Yes, making and eating dumplings with friends was a lot of fun.
是的，和朋友们一起做饺子、吃饺子，真是很有趣。

Funny facts 关于饺子的有趣事实和短语

节日习俗：饺子是中国传统节日，尤其是春节的重要食品，北方人常在除夕夜吃饺子，寓意团圆和吉祥。

吃法多样：饺子可以煮、蒸、煎、炸，煮饺子是最常见的吃法，但煎饺和炸饺子也深受喜爱。

寓意丰富：在饺子里包硬币、干果、糖果等是传统习俗，象征好运、甜蜜和福气，吃到这些饺子的人被认为会有好运。

Lunar New Year's Eve – 除夕夜
stuffing/filling – 馅料
dumpling wrapper – 饺子皮
boiled dumplings – 煮饺子

steamed dumplings – 蒸饺
pan-fried dumplings – 煎饺
handmade dumplings – 手工饺子

Writing practice 写作小练习

根据我们这一节所学到的内容，写出以下句子的英文。

1. 包饺子是非常有趣的。

   _____

2. 饺子的馅料可以是肉、蔬菜或海鲜。

   _____

3. 饺子的形状看起来像一艘小船。

   _____

Reference translation 参考译文

每当中国新年到来时，家家户户都会聚在一起庆祝，而其中一道必不可少的美食就是饺子。让我们一起来探索这种深受喜爱的中国传统美食的故事、传统和风味吧！

饺子，是中国传统的主食，特别是在春节期间。这些美味的食物由薄薄的面皮包裹着各种馅料制成，如肉末、蔬菜和香料。它们可以煮、蒸或煎，做得恰到好处。

饺子的起源有很多种说法，其中一个有趣的说法是，饺子是1,800多年前由一位著名的中医张仲景发明的。他制作饺子来帮助人们在严寒的冬季抵御寒冷和疾病。在春节期间吃饺子有着悠久的历史，饺子的形状像古代的银元宝，象征着财富和繁荣，因此它们是迎接新年的完美佳肴。

准备饺子是一项充满欢乐的家庭活动，通常会动员全家人参与。除夕夜，家人们围坐在桌旁，一起制作饺子，每个人都贡献一份力量。面团被擀成薄薄的圆片，馅料准备好后，每个饺子被小心地包好并封口。有些家庭甚至会在一个饺子里藏一枚硬币——谁吃到它，来年就会好运连连！

饺子具有极好的口感。面皮柔软而略带嚼劲，馅料鲜美可口。常见的馅料包括猪肉白菜馅、虾仁韭菜馅以及芥菜等素食选项。当你咬上一口新鲜出锅的饺子时，丰富的味道会在口中爆发，带来一场美妙的味觉盛宴。

饺子通常与蘸料一同上桌，蘸料由酱油、醋，有时还加入一点辣椒油或蒜末制成。这为饺子增添了额外的风味，使每一口都更加美味。无论是煮的、蒸的还是煎的，饺子都是一道百搭且备受欢迎的美食，为每一餐带来温暖与幸福。

在春节期间，饺子不仅仅是一种食物，它们还是家庭团聚和欢庆的象征。制作和品尝饺子是庆祝新年、传递美好祝愿以及与亲人共创难忘回忆的方式。

Laba Porridge
腊八粥：腊八节的传统美食

Listening & practice 听英文原声，完成练习

▶扫码听音频◀

1. **When is Laba Porridge traditionally eaten?**
   A. On the eighth day of the twelfth month of the lunar calendar
   B. On the first day of the first month of the lunar calendar
   C. On the fifteenth day of the eighth month of the lunar calendar

2. **What is Laba Porridge made from?**
   A. Meat and vegetables
   B. A variety of grains, beans, nuts, and dried fruits
   C. Fish and noodles

3. **What does Laba Porridge commemorate?**
   A. The Chinese New Year
   B. The birth of Confucius
   C. The enlightenment of the Buddha

4. **What is another name for Laba Porridge?**
   A. Buddha Porridge
   B. Dragon Boat Porridge
   C. Mooncake Porridge

5. **Why do people make extra Laba Porridge on the Laba Festival?**
   A. To sell at markets
   B. To use as decorations
   C. To share with friends, neighbours, and the less fortunate

Reading 阅读下面的文章

In China, as winter approaches and the weather turns colder, there is a special festival that warms everyone's hearts and stomachs—the Laba Festival. One of the highlights of this festival is Laba Porridge.

Laba Porridge is a traditional Chinese porridge made from a variety of grains, beans, nuts, and dried fruits. It is typically enjoyed on the eighth day of the twelfth month of the lunar calendar, known as the Laba Festival. This festival marks the beginning of preparations for the Spring Festival and holds rich historical and cultural significance.

The origin of Laba Porridge can be traced back over a thousand years to the Song Dynasty. According to a widely known legend, this festival commemorates the enlightenment of the Buddha. It is said that after years of meditation and hardship, the Buddha attained enlightenment on this day. To celebrate this event, people began cooking a rich and hearty porridge made from stored grains and beans. Therefore, Laba Porridge is also known as "Buddha Porridge".

Preparing Laba Porridge is a joyful and communal family activity. Families gather in the kitchen to combine a variety of ingredients, including rice, millet, glutinous rice, red beans, mung beans, peanuts, lotus seeds, jujubes, and dried fruits. Each ingredient adds its own unique flavour and nutritional value to the porridge. The mixture is simmered slowly for several hours until it becomes thick and creamy.

Laba Porridge has a rich flavour, perfectly balancing the sweetness of dried fruits with the slight nuttiness of peanuts and lotus seeds. The combination of different grains and beans creates a satisfying texture that warms you from the inside out.

Laba Porridge is more than just a dish; it's a tradition that brings families together. On the Laba Festival, people often make extra porridge to share with friends and neighbours, spreading warmth and kindness throughout the community. Some families even prepare large pots of Laba Porridge to give to the less fortunate, embodying the spirit of generosity and compassion.

In addition to its delicious taste and cultural significance, Laba Porridge is also highly nutritious. It is packed with vitamins, minerals, and protein from the diverse ingredients, making it a wholesome and balanced meal. This makes it especially perfect for the cold winter months when people need extra **nourishment** to stay healthy.

Vocabulary and phrases 词汇和短语

approach [ə'prəʊtʃ] 动 靠近；接近

typically ['tɪpɪkli] 副 典型地；代表性地

meditation [ˌmedɪ'teɪʃn] 名 沉思；冥想

nuttiness ['nʌtɪnəs] 名 果仁味

community [kə'mjuːnəti] 名 社区；群体

compassion [kəm'pæʃn] 名 怜悯；同情

porridge ['pɒrɪdʒ] 名 粥

enlightenment [ɪn'laɪtnmənt] 名 启迪；启开；开明

sweetness ['swiːtnəs] 名 甜美

satisfying ['sætɪsfaɪɪŋ] 形 令人满意的；圆满的

generosity [ˌdʒenə'rɒsəti] 名 慷慨；大方

nourishment ['nʌrɪʃmənt] 名 营养；滋养品

Practice 请选择合适的词填在下方的横线上

approaches   Porridge   typically   satisfying   community

1. Laba _____ is full of different nuts and beans.
2. In our _____, everyone enjoys sharing Laba Porridge with their neighbours during the festival.

3. Laba Porridge is _____ made with a variety of ingredients like rice, beans, nuts, and dried fruits.

4. The variety of ingredients in Laba Porridge makes it a _____ and tasty meal.

5. As the day of Laba _____, families start preparing delicious Laba Porridge.

Talking practice 情景对话模拟练习

腊八节的时候，孟硕请比尔品尝了腊八粥，让我们跟着这段对话进行练习吧。

Bill, try this porridge. It's called Laba Porridge.
比尔，尝尝这碗粥，它叫腊八粥。

Meng Shuo 孟硕

This porridge tastes really good! How is it made?
这个粥很好喝啊，它是怎么做的？

Bill 比尔

We use many ingredients like rice, beans, nuts, and dried fruits.
我们用很多食材，比如大米、豆子、坚果和干果。

Meng Shuo 孟硕

Do you often drink this porridge?
你们经常喝这种粥吗？

Bill 比尔

We eat it during the Laba Festival, which is on the eighth day of the twelfth month of the lunar calendar.
我们在腊八节喝，这个节日在农历十二月的第八天。

Meng Shuo 孟硕

What is special about the Laba Festival?
腊八节有什么特别之处？

Bill 比尔

It marks the beginning of preparations for the Spring Festival.
它标志着春节准备的开始。

Meng Shuo孟硕

That's interesting! Do your families make it together?
真有趣！家人会一起做吗？

Bill比尔

Yes, it's a family activity. We also share it with friends and neighbours.
是的，这是一个家庭活动。我们还会和朋友和邻居分享。

Meng Shuo孟硕

Funny facts 关于腊八粥的有趣事实和短语

营养丰富：腊八粥由多种谷物和豆类混合煮制而成，富含蛋白质、纤维素和多种维生素，是一种非常营养的食品。

养生食品：腊八粥中的多种谷物和豆类具有补气血、健脾胃、暖身驱寒的功效，是冬季养生的佳品。

the Laba Festival – 腊八节
mixed grains – 杂粮
red beans – 红豆
mung beans – 绿豆

lotus seeds – 莲子
jujubes – 枣
walnuts – 核桃
chestnuts – 栗子

Writing practice 写作小练习

根据我们这一节所学到的内容，写出以下句子的英文。

1. 腊八粥能让我们在寒冷的冬天里保持温暖和营养。

   _____

2. 腊八粥是由多种谷物、豆类、坚果和干果制成的。

   _____

3. 腊八粥味道甜美可口。

Reference translation 参考译文

在中国，随着冬季的到来，天气逐渐变冷，但有一个特别的节日可以温暖每个人的心和胃，这就是腊八节。这个节日的亮点之一就是腊八粥。

腊八粥是中国传统粥品，由多种谷物、豆类、坚果和干果制成。通常在农历十二月初八，也就是腊八节这一天享用。这个节日标志着春节准备工作的开始，并承载着丰富的历史和文化意义。

腊八粥的起源可以追溯到一千多年前的宋代。有一个广为传颂的说法是，这个节日是为了纪念佛祖成道。据传说，经过多年的冥想和艰苦修行，佛祖在这一天达到了悟道之境。为了庆祝这一事件，人们开始用储存的谷物和豆类烹制一碗丰富且滋补的粥。所以腊八粥也被称为"佛粥"。

熬制腊八粥是一项充满欢乐的家庭活动。家人们聚在厨房里，将各种食材混合在一起，包括大米、小米、糯米、红豆、绿豆、花生、莲子、枣和干果。每种食材都为粥增添了独特的风味和营养价值。将这些混合物慢慢炖煮数小时，直到粥变得浓稠而香甜。

腊八粥味道丰富，完美平衡了干果的甜味与花生和莲子的微微坚果香。不同谷物和豆类的结合创造出一种令人满足的口感，从内到外温暖着你的身心。

腊八粥不仅仅是一道美食，它也是一种将家庭团结在一起的传统。在腊八节，人们常常做多一些腊八粥与朋友和邻居分享，将温暖和善意传播到整个社区。有些家庭甚至准备大锅的腊八粥，赠送给那些不幸的人，体现了慷慨与同情的精神。

除了美味和文化意义之外，腊八粥还极富营养。它由多种食材组成，富含维生素、矿物质和蛋白质，是一顿全面而均衡的餐食。这使得腊八粥在寒冷的冬季尤其完美，因为人们需要额外的营养来保持健康。

Niangao
年糕：新年步步高升的象征

Listening & practice 听英文原声，完成练习

▶扫码听音频◀

1. What is another name for niangao?
   A. New Year Cake
   B. Moon Cake
   C. Spring Cake

2. What does the name "niangao" symbolize?
   A. Growth and progress
   B. Wealth and power
   C. Health and happiness

3. What are the main ingredients of niangao?
   A. Glutinous rice flour, water, and sugar
   B. Wheat flour, water, and honey
   C. Cornmeal, milk, and butter

4. How is niangao typically prepared?
   A. Steamed until firm and sticky
   B. Baked in an oven
   C. Boiled in water

5. Why do families share niangao during the Chinese New Year?
   A. To welcome new friends
   B. To symbolize wishes for good luck and prosperity
   C. To celebrate birthdays

Reading 阅读下面的文章

Besides dumplings, there is another special treat that is highly popular during the Chinese Lunar New Year—niangao (rice cake), or New Year Cake. This delightful dish brings **endless** joy and **deliciousness** to the celebrations.

Niangao is a traditional Chinese rice cake made from glutinous rice flour and sugar. It is usually enjoyed during the Spring Festival and is loved for its sticky texture and sweet flavour. The name "niangao" sounds like "year higher" in Chinese, symbolizing growth, progress, and the promise of a better year ahead.

Niangao has a long history and is considered one of the oldest festive foods in China. A recipe for niangao was **mentioned** in a cookbook from the 6th century, showing its **deep-rooted** tradition.

The basic ingredients of niangao include glutinous rice flour, water, and sugar. To **enhance** the flavour and texture, people often add Chinese red dates, nuts, or sweet beans. The mixture is **poured** into a **mold** and steamed until it becomes firm and sticky.

The texture of niangao is satisfyingly chewy, and its sweet flavour is often **accompanied** by the aroma of fruits or nuts, making it a delicious treat. There are many ways to enjoy niangao: it can be steamed, fried, or even added to savory dishes. Some people like to slice and pan-fry niangao until it becomes crispy on the outside and soft on the inside, creating a perfect blend of textures.

Niangao is more than just a cake; it carries deep cultural significance. During the Chinese New Year, families gather to share niangao, symbolizing their wishes for good luck and prosperity. The sticky texture of the cake represents unity and family **cohesion**, while its name signifies good fortune and happiness. It is also **common** to give niangao as gifts to friends and neighbours, spreading joy and well-wishes.

In addition to its cultural significance, niangao is also highly versatile and nutritious. It provides energy and warmth during the cold winter months, making it a perfect food for the New Year celebrations. Its simple ingredients make it easy to prepare, and its delightful taste **ensures** that it is loved by both young and old.

Vocabulary and phrases 词汇和短语

endless ['endləs] 形 无止境的	deliciousness [dɪ'lɪʃəsnɪs] 名 美味；好吃
mention ['menʃn] 动 提到；说起	deep-rooted [ˌdiːp 'ruːtɪd] 形 根深蒂固的
enhance [ɪn'hɑːns] 动 提高；增加	pour [pɔː(r)] 动 倒；倾泻
mold [məʊld] 名 模子；模型	accompany [ə'kʌmpəni] 动 陪伴；伴随……发生
cohesion [kəʊ'hiːʒn] 名 凝聚力；团结	common ['kɒmən] 形 常见的；普通的
ensure [ɪn'ʃʊə(r)] 动 保证；确保	

Practice 请选择合适的词填在下方的横线上

> endless   mentioned   accompanies   common   ensure

1. When we talk about Chinese New Year, niangao is always _____ as a must-eat food.

2. Niangao is a _____ traditional dessert that people eat to celebrate the new year.

3. Eating niangao brings _____ happiness and good luck to our family.

4. The tradition of eating niangao _____ us from generation to generation.

5. Eating niangao during the Spring Festival is believed to _____ good luck and prosperity for the coming year.

Talking practice 情景对话模拟练习

安迪对中国新年的传统美食很感兴趣，刘丹给他讲解了年糕，你也跟着练习对话中的内容吧。

Chinese New Year is so interesting! Liu Dan, besides dumplings, what other traditional foods do you have for the New Year?
中国的新年真是太有意思了！对了，刘丹，除了饺子以外，你们新年还有哪些传统美食？

Andy 安迪

Liu Dan 刘丹

We also have niangao, or New Year Cake. It's a sweet rice cake.
我们还有年糕，也叫新年糕。它是一种甜米糕。

niangao? What's it made of?
年糕？它是用什么做的？

Andy 安迪

Liu Dan 刘丹

It's made from glutinous rice flour and sugar. Sometimes we add nuts or sweet beans.
它是用糯米粉和糖做的。有时候我们还会加坚果或甜豆。

How do you eat niangao?
你们怎么吃年糕？

Andy 安迪

Liu Dan 刘丹

We can steam it, fry it, or even add it to savory dishes. It's very versatile!
我们可以蒸着吃，煎着吃，甚至可以加到咸的菜里。它非常多样化！

What does niangao symbolize?
年糕象征着什么？

Andy 安迪

年糕：新年步步高升的象征　147

niangao symbolizes growth and progress. Its name sounds like 'year higher' in Chinese.
年糕象征着成长和进步。它的名字在中文里听起来像'年年高'。

Liu Dan 刘丹

That sounds delicious and meaningful!
听起来既美味又有意义！

Andy 安迪

Funny facts　关于年糕的有趣事实和短语

主要食材：年糕的主要成分是糯米或粳米，有的地方还加入红枣、红豆、桂花等，增加风味和营养。

食用方法：年糕的食用方法多种多样，可以蒸、煮、煎、炒、炸，常见的有红烧年糕、年糕汤、年糕火锅等。

寓意吉祥："年糕"在中文中与"年高"同音，寓意着年年高升、步步高升，象征着人们对美好生活的向往。

---

sweet rice cake – 甜年糕
savory rice cake – 咸年糕
steamed rice cake – 蒸年糕

fried rice cake – 炸年糕
stir-fried rice cake – 炒年糕
sticky texture – 黏糯口感

Writing practice　写作小练习

根据我们这一节所学到的内容，写出以下句子的英文。

1. 年糕是用糯米粉做的，这让它有黏黏的、有嚼劲的口感。

　

2. 吃年糕象征着新年好运和繁荣。

　

3. 人们经常在年糕中加入红豆、坚果或其他食材。

Reference translation 参考译文

除了饺子以外,在中国农历新年期间,还有一种特别的美食也颇受人们的欢迎,这就是年糕,这道美味佳肴为庆祝活动增添了无尽的欢乐和美味。

年糕,是用糯米粉和糖制成的中国传统米糕。它通常在春节期间食用,因其黏糯的口感和甜美的风味而深受喜爱。"年糕"这个名字在中文中听起来像"年年高升",象征着成长、进步和来年更加美好的愿望。

年糕拥有悠久的历史,被视为中国最古老的节庆食品之一。早在六世纪的食谱中就有年糕的食谱记载,显示出其深厚的传统底蕴。

年糕的基本原料包括糯米粉、水和糖。为了增加风味和口感,人们还会加入红枣、坚果或甜豆等食材。将混合物倒入模具中蒸煮,直到它变得坚实而黏糯。

年糕的口感满足而富有弹性,甜美的风味常常伴随着水果或坚果的香气,使其成为一道美味佳肴。年糕有多种吃法:可以蒸、煎,甚至加入咸味的菜肴中。有些人喜欢将年糕切片煎至外脆里嫩,创造出一种完美的口感组合。

年糕不仅仅是一种糕点,它还承载着深厚的文化意义。在农历新年期间,家人们聚在一起分享年糕,寓意互相祝愿好运和繁荣。年糕的黏糯质地象征着团结和家庭的凝聚力,它的名字寓意着吉祥和好运。人们还常常将年糕作为礼物送给朋友和邻居,传播快乐和美好的祝愿。

除了文化意义,年糕也非常多样化和营养丰富。在寒冷的冬季,它提供了能量和温暖,是新年庆祝活动的完美食品。其简单的原料使得每个人都能轻松制作,而其美妙的味道确保了它深受年轻人和老年人的喜爱。

Part 4
独具特色的中国饮食文化

Regional Food Cultures in China
中国的地方饮食文化：风味之旅

Listening & practice 听英文原声，完成练习

▶扫码听音频◀

1. What flavour is Yunnan cuisine known for?
   A. Spicy
   B. Sweet
   C. Sour

2. What makes Sichuan cuisine unique?
   A. Sweet flavours
   B. Salty flavours
   C. Spicy and numbing flavours

3. What flavour is common in Northeast China (Dongbei) cuisine?
   A. Sweet
   B. Salty
   C. Bitter

4. How is Cantonese cuisine often cooked?
   A. Boiled
   B. Roasted
   C. Steamed and stir-fried

5. What flavour is Jiangsu cuisine known for?
   A. Sweet
   B. Spicy
   C. Sour

Reading 阅读下面的文章

China is a vast country with a rich and diverse culinary heritage. Different regions have unique flavours and cooking styles that make their food special.

In Yunnan, people love sour flavours. The mountainous region of Yunnan is home to many ethnic groups, each with its own distinct cuisine. One of the most popular ingredients in Yunnanese cooking is pickled vegetables, which add a tangy taste to many dishes. The famous "Crossing-Bridge Rice Noodles" often includes sour pickled vegetables, making the soup bright and flavourful. Yunnan's love for sour food is a delightful and refreshing aspect of their cuisine.

In Sichuan, people are known for their love of spicy food. Sichuan cuisine is famous for its bold and spicy flavours, often created with chili peppers and Sichuan peppercorns. The peppercorns give a unique numbing sensation that enhances the spiciness. Dishes like "Mapo Tofu" and "Kung Pao Chicken" are perfect examples of Sichuan's fiery taste. If you enjoy a little heat in your food, Sichuan cuisine is sure to delight your taste buds!

In Northeast China, or Dongbei, people prefer salty flavours. The cold climate in this region has influenced the local cuisine to be hearty and savory. Pickling and salting are common preservation methods, resulting in dishes that are rich in salt and flavour. "Chicken and Mushroom Stew" and "Braised Pork with Vermicelli" are popular dishes that showcase the region's love for salty food. The robust and comforting flavours of Dongbei cuisine are perfect for warming up during the chilly winters.

In Guangdong, also known as Canton, people enjoy light and fresh flavours. Cantonese cuisine focuses on preserving the natural taste of the ingredients. Steaming and stir-frying are common cooking methods that help maintain the freshness of the food. Dim sum, a traditional Cantonese meal, includes a variety of small dishes like steamed dumplings and buns, each bursting with delicate flavours. If you appreciate subtle and fresh tastes, Cantonese cuisine will surely impress you.

In Jiangsu, people favour sweet flavours. Jiangsu cuisine, also known as Su cuisine, is known for its use of sugar to create a balance of sweet and savory flavours. Dishes like "Sweet and Sour Mandarin Fish" and "Braised Pork" are cooked with a touch of sweetness, making them rich and delicious. The intricate balance of flavours in Jiangsu cuisine is a treat for anyone with a sweet tooth.

Each region in China has its own unique food culture that reflects the local climate, geography, and traditions. From the sour dishes of Yunnan to the spicy delights of Sichuan, the salty flavours of Dongbei to the fresh tastes of Canton, and the sweet treats of Jiangsu, exploring Chinese regional cuisines is like embarking on a flavourful adventure.

Vocabulary and phrases 词汇和短语

diverse [daɪ'vɜːs] 形 不同的；多种多样的

region ['riːdʒən] 名 地区

pickled ['pɪkld] 形 腌制的；腌渍的

spiciness ['spaɪsɪnəs] 名 香郁；富于香料

preservation [ˌprezə'veɪʃn] 名 保存；维护

geography [dʒi'ɒgrəfi] 名 地理

mountainous ['maʊntənəs] 形 多山的

ethnic ['eθnɪk] 形 种族的；民族的

numbing ['nʌmɪŋ] 形 使麻木的；使失去感觉的

influence ['ɪnfluəns] 动 影响

appreciate [ə'priːʃieɪt] 动 欣赏；喜爱

embarking [ɪm'bɑːk] 名 乘船；从事

Practice 请选择合适的词填在下方的横线上

ethnic   numbing   Influenced   appreciate   geography

1. Eating hot pot in Sichuan is a must-try experience, with its _____ broth and delicious ingredients.

中国的地方饮食文化：风味之旅

2. _____ by climate, each region in China possesses its own unique food culture.

3. I really _____ the delicious dim sums and seafood in Cantonese cuisine.

4. China's food culture reflects the local climate, _____, and traditions.

5. Yunnan's food culture is famous for its rich _____ flavours.

Talking practice 情景对话模拟练习

苏菲亚对中国不同地区的美食风格很感兴趣，高珊为她做了介绍，你也跟着练习这段对话吧。

> Gao Shan, China is so big. Are there many different food styles in different places?
> 高珊，中国这么大，是不是不同的地方美食风格有很大区别呢？

Sophia苏菲亚

Gao Shan高珊

> Yes, there are. For example, in Yunnan, people love sour flavours.
> 是的，有很大的区别。比如，在云南，人们喜欢酸味。

> What about other places?
> 那其他地方呢？

Sophia苏菲亚

Gao Shan高珊

> In Sichuan, they love spicy food. It's very hot!
> 在四川，人们喜欢吃辣的食物，非常辣！

> That sounds exciting. What about in the north?
> 听起来很刺激。那北方呢？

Sophia苏菲亚

Gao Shan高珊

> In Northeast China, the food is salty. They use a lot of pickles and salted meats.
> 在中国东北，食物很咸。他们多用酸菜和腌肉。

What is the food like in southern China?
中国南方的食物怎么样?

Sophia 苏菲亚

Gao Shan 高珊

In Guangdong, they like light and fresh flavours. Dim sum is very famous there.
在广东，人们喜欢清淡和新鲜的味道。那里的点心非常有名。

I love trying different foods! Chinese cuisine sounds amazing.
我喜欢尝试不同的食物！中国菜听起来很棒。

Sophia 苏菲亚

Funny facts 关于中国地方饮食文化的有趣事实和短语

八大菜系：鲁菜、川菜、粤菜、苏菜、浙菜、闽菜、湘菜和徽菜。每个菜系都有独特的烹饪方法和风味特点。

主食差异：中国北方以面食为主，如北京的炸酱面、山西的刀削面和陕西的油泼面，而南方则以米饭为主食。

酱料多样：中国各地有丰富多样的酱料，如北京的甜面酱、四川的豆瓣酱、广东的蚝油和湖南的剁椒酱，这些酱料为当地菜肴增色不少。

---

eight major cuisines – 八大菜系
noodle culture – 面食文化
hot pot culture – 火锅文化

dim sum culture – 点心文化
local snacks – 风味小吃

Writing practice 写作小练习

根据我们这一节所学到的内容，写出以下句子的英文。

1. 广东人喜欢喝早茶，品尝各种点心。

   _____

2. 中国不同的地方有自己特色的美食。

   _____

3. 在中国，饮食文化代表着一种传统和习俗。

Reference translation 参考译文

中国是一个拥有丰富多样饮食文化的广袤国家。不同的地区有独特的风味和烹饪风格，使其食物独具特色。

在云南，人们喜欢酸味。这个多山的地区是众多少数民族的家园，每个民族都有自己独特的美食。云南菜中最为流行的食材之一是腌菜，为许多菜肴增添了酸爽的味道。著名的"过桥米线"常常包括腌菜，使得汤底鲜美且风味十足。云南人对酸味食品的喜爱，为其菜肴增添了一份令人愉悦的清新感。

在四川，人们以爱吃辣而闻名。川菜因其大胆和辛辣的味道而著称，这种风味通常由辣椒和四川花椒调制而成。花椒带来一种独特的麻感，使得辣味更加突出。像"麻婆豆腐"和"宫保鸡丁"这样的菜肴完美体现了四川的火辣风味。如果你喜欢在食物中感受一点热辣，川菜一定会让你的味蕾大为满足！

在中国东北，人们偏爱咸味。这个地区的寒冷气候影响了当地的饮食，使其变得浓郁而美味。腌制和盐渍是常见的保存方法，因此菜肴中富含盐分和风味。"小鸡炖蘑菇"和"猪肉炖粉条"是展示东北人对咸味喜爱的热门菜肴。东北菜肴的浓郁和舒适感是寒冷冬季暖身的完美选择。

在广东，人们喜欢清淡和新鲜的味道。粤菜注重保持食材的原汁原味。蒸制和快炒是常见的烹饪手法，有助于保持食物的鲜美。点心是传统的粤式餐点，包括各种小菜，如蒸饺和包子，充满了精致的味道。如果你钟爱淡雅清新的口感，粤菜定会令你赞叹不已。

在江苏，人们喜欢甜味。江苏菜，也被称为苏菜，以使用糖来创造甜咸平衡的风味而闻名。"松鼠鳜鱼"和"红烧肉"是用糖调味的代表菜肴，使它们口感丰富而美味。江苏菜精妙的风味平衡，对任何喜欢甜食的人来说都是一场盛宴。

中国的每个地区都有其独特的饮食文化，反映了当地的气候、地理和传统。从云南的酸味菜肴到四川的辛辣美味，从东北的咸香食物到广东的清鲜口味，再到江苏的甜美佳肴，探索中国各地的美食就像踏上了一场味蕾的冒险之旅。

Chinese Cooking Techniques
中国的烹饪技巧：中餐美味的秘诀

Listening & practice 听英文原声，完成练习

▶扫码听音频◀

1. What cooking method uses a wok to cook food quickly at high heat?
   A. Stir-frying
   B. Steaming
   C. Boiling

2. Which cooking technique is often used to make soups and noodles?
   A. Stir-frying
   B. Boiling
   C. Deep-frying

3. Which cooking method uses bamboo steamers to cook food gently?
   A. Deep-frying
   B. Boiling
   C. Steaming

4. How is food cooked using the braising (烧) technique?
   A. Cooked quickly at high heat
   B. Cooked in a large amount of liquid over high heat
   C. Cooked slowly in a small amount of liquid over low heat

5. What famous Chinese dish is made by roasting?
   A. Beijing duck
   B. Kung Pao Chicken
   C. Fried rice

Reading 阅读下面的文章

Have you ever wondered how Chinese food gets its amazing flavours and textures? Let's learn about some fun and interesting Chinese cooking techniques!

Stir-frying is one of the most popular Chinese cooking methods. It involves cooking food quickly at high heat in a wok. Ingredients like vegetables, meat, and tofu are cut into small pieces and cooked in hot oil with various seasonings. This method makes the food bright, colourful, and crispy. Stir-frying is perfect for dishes like Fried Rice and Kung Pao Chicken.

Steaming is another important Chinese cooking technique. Food is placed in bamboo steamers, which are stacked on top of each other and set over boiling water. This gentle cooking method helps keep the food's natural flavours and nutrients. Steamed dumplings and buns, often seen in dim sum, are popular steamed dishes. Steamed fish and vegetables are also common.

Boiling is a simple and quick way to cook food. Food is cooked in boiling water or broth. This method is used for making soups, noodles, and hot pots. In hot pot cooking, a pot of boiling broth is placed in the centre of the table, and diners cook their own ingredients by dipping them into the hot broth. It's a fun and interactive way to enjoy fresh and tender food.

Braising involves cooking food slowly in a small amount of liquid over low heat. This technique makes the food flavourful and tender. Dishes like Braised Pork and Braised Beef are made this way. The food is first browned in oil, then simmered in a savory sauce made from soy sauce, sugar, and spices. Braising allows the flavours to fully develop.

Deep-frying is used to make food crispy and golden. Food is submerged in hot oil until it becomes crispy on the outside and cooked through on the inside. Spring Rolls, Sweet and Sour Chicken, and Sesame Balls are popular deep-fried dishes. This method gives food a delightful crunch and a tasty flavour.

Roasting involves cooking food in an oven or over an open flame. Beijing roast duck is a famous Chinese dish that uses **roasting**. The duck is seasoned and hung to dry before being roasted until the skin is crispy and the meat is tender. Roasting is also used for making char siu (barbecued pork) and roasted meats, known for their smoky flavours and juicy textures.

Chinese cooking techniques are as diverse as the dishes they create. Each method **brings out** unique flavours and textures, making Chinese cuisine rich and **varied**. Whether you're stir-frying vegetables, steaming dumplings, or roasting duck, these techniques ensure that every meal is delicious and enjoyable.

Vocabulary and phrases 词汇和短语

technique [tek'niːk] 名 技术；技巧

stack [stæk] 动 堆积；堆放

braising ['breɪzɪŋ] 名 炖；烧

submerge [səb'mɜːdʒ] 动 使浸没

roasting ['rəʊstɪŋ] 名 烧烤

varied ['veərid] 形 各种各样的

steamer ['stiːmə(r)] 名 蒸笼

interactive [ˌɪntər'æktɪv] 形 相互作用的；交互的

develop [dɪ'veləp] 动 发展

crunch [krʌntʃ] 名 嘎吱声

bring out 短 推出；拿出

Practice 请选择合适的词填在下方的横线上

> technique   stacked   develop   bring out   varied

1. Today, Chinese cooking techniques continue to _____.
2. In steaming, ingredients are _____ in bamboo steamer baskets.

中国的烹饪技巧：中餐美味的秘诀

3. Chinese cooking techniques _____ the natural flavours of ingredients.

4. Steaming is an important cooking _____ in Chinese cuisine.

5. The diverse cooking techniques of China make Chinese cuisine rich and _____.

Talking practice 情景对话模拟练习

丹尼斯非常好奇美味的中国菜是如何制作的，让我们跟着郭鹏的回答来练习相关的对话吧。

Guo Peng, why are there so many different types of Chinese food, and why do they all taste so good?
郭鹏，为什么中国菜种类这么多，而且都很好吃呢？

Dennis丹尼斯

Guo Peng郭鹏
It's because of the different cooking techniques we use.
这是因为我们用不同的烹饪方法。

What kind of techniques?
有哪些方法呢？

Dennis丹尼斯

Guo Peng郭鹏
Well, we have stir-frying, steaming, boiling, braising, deep-frying, and roasting.
嗯，我们有炒、蒸、煮、烧、炸和烤这些方法。

Can you tell me about stir-frying?
你能告诉我炒菜吗？

Dennis丹尼斯

Guo Peng郭鹏
Sure! Stir-frying cooks food quickly at high heat in a wok. It's great for making Fried Rice and Kung Pao Chicken.
当然！炒菜是在高温下快速烹饪食物，非常适合做炒饭和宫保鸡丁。

Dennis丹尼斯

What about steaming?
那蒸菜呢？

Guo Peng郭鹏

Steaming uses bamboo steamers over boiling water. It keeps the food's natural flavours. We steam dumplings, buns, and fish.
蒸菜是用竹蒸笼在开水上蒸。这可以保持食物的原味。我们蒸饺子、包子和鱼。

Dennis丹尼斯

I see! No wonder Chinese food is so delicious.
我明白了！难怪中国菜这么好吃。

Funny facts 关于中餐烹饪方法的有趣事实和短语

卤：卤是将食材放入卤水中煮熟，使其入味，常用于卤蛋、卤肉和卤豆腐。
凉拌：凉拌是将食材切好后加入调料凉拌，常用于凉拌黄瓜、凉拌木耳和凉拌皮蛋。
红烧：红烧是将食材在糖和酱油中慢火煮至酥烂，常用于红烧肉、红烧鱼和红烧排骨。

---

pan-frying – 煎
marinating – 卤

cold mixing – 凉拌

Writing practice 写作小练习

根据我们这一节所学到的内容，写出以下句子的英文。

1. 蒸是一种健康的烹饪方式，可以保留食物的原汁原味。

_____

2. 煮很简单，你只需要把食物放在水里煮到熟就可以了。

_____

3. 烤可以让食物的外皮变得酥脆，而内部保持柔软多汁。

Reference translation 参考译文

你是否曾经好奇过中国菜是如何拥有如此令人惊艳的口味和口感的？让我们来了解一些有趣的中国烹饪技艺吧！

炒是最受欢迎的中国烹饪方法之一。这种方法是将食材在高温下快速翻炒。蔬菜、肉类和豆腐等食材被切成小块，在热油中与各种调味料一起翻炒。这种方法使食物色彩鲜艳，口感酥脆。炒非常适合制作炒饭和宫保鸡丁等菜肴。

蒸是另一种重要的中国烹饪技艺。食物被放置在竹蒸笼中，这些蒸笼堆叠在一起，然后置于沸水上。这种温和的烹饪方法有助于保留食物的天然风味和营养。蒸饺和包子，这些在点心店常见的食物，都是流行的蒸制食物。蒸鱼和蔬菜是常见的蒸菜。

煮是一种简单快捷的烹饪方法。食物被放在沸水或高汤中煮熟。这种方法用于制作汤、面条和火锅。吃火锅时，一锅沸腾的高汤被放在餐桌中央，食客们将各种食材放入热汤中煮熟。这是一种享受新鲜嫩滑食物的有趣且互动的方式。

烧是一种将食物在少量液体中慢慢炖煮的方法。这种方法使食物味道浓郁且嫩滑。红烧肉和红烧牛肉就是用这种方法制作的。食物先在油中煎至棕褐色，然后在由酱油、糖和香料制成的调料中慢炖。烧可以充分发展食材的味道。

炸是一种使食物酥脆金黄的烹饪方法。食物被完全浸入热油中，直到外皮变得酥脆，内部熟透。春卷、糖醋鸡和芝麻球是很受欢迎的油炸食物。这种烹饪方法赋予了食物令人愉悦的酥脆口感和美味的滋味。

烤是将食物放在烤箱或明火上烤制。北京烤鸭是一道著名的中国菜肴，便采用了烘烤的手法。鸭子先被腌制，然后挂起晾干，再烘烤至外皮酥脆、肉质鲜嫩。烘烤还用于制作叉烧（烧烤猪肉）和烤肉，这些菜肴以烟熏风味和多汁口感而著称。

中国烹饪技艺和它们制作的菜肴一样丰富多样。每种方法都能带出独特的风味和口感，使中国菜肴丰富多彩。无论你是做炒菜、蒸饺子还是烤鸭，这些技艺都确保每顿饭都美味可口。

Chinese Table Manners
中国的餐桌礼仪：吃饭的文化

Listening & practice 听英文原声，完成练习

▶扫码听音频◀

1. What is not polite to do with chopsticks?
   A. Stick them upright in a bowl of rice
   B. Use them to pick up food gently
   C. Place them beside your plate

2. Who usually sits facing the entrance at a Chinese dining table?
   A. The most honoured guest or the eldest person
   B. The youngest person
   C. The host

3. What should you do before serving yourself?
   A. Serve the most expensive dish
   B. Serve others first
   C. Take the biggest portion

4. How are meals often served in China?
   A. Family-style
   B. Buffet-style
   C. Individually plated for each person

5. When is it polite to start eating?
   A. After the host or the eldest person starts
   B. As soon as you sit down
   C. After finishing your drink

Reading 阅读下面的文章

Chinese table manners are an important part of Chinese culture. When you eat with others, it's not just about the food—it's also about showing respect and being polite. Let's learn some fun and interesting facts about Chinese table manners!

In China, people use chopsticks to eat most of their food. When using chopsticks, there are a few rules to remember. For example, don't stick your chopsticks upright in a bowl of rice because it looks like incense sticks at a funeral, which is considered bad luck. Also, don't point with your chopsticks or play with them. Use them to pick up food gently and place it into your mouth.

Seating arrangements are important at a Chinese dining table. The seat facing the entrance is usually reserved for the most honoured guest or the eldest person. Younger people or those with lower status sit closer to the entrance. This shows respect to the elders and the most important guests.

In China, it's polite to serve others before serving yourself. This means you should use the serving spoons or chopsticks to put food on others' plates before taking your own portion. This shows that you care about others and are thoughtful.

Chinese meals are often served "family-style", with large dishes placed in the centre of the table for everyone to share. This way, everyone can try a little bit of everything. It's important to take only what you can eat and not waste food. Sharing food also symbolizes unity and togetherness.

When dining with others, it's polite to wait for the host or the eldest person to start eating first. This shows respect and good manners. It's also customary to say "let's eat" or "enjoy your meal" before everyone starts eating.

Chinese table manners are about showing respect, being polite, and enjoying the meal together. By following these simple rules, you can show that you appreciate the food and the company. Next time you have a meal, remember these Chinese table manners and enjoy your dining experience!

Vocabulary and phrases 词汇和短语

manner ['mænə(r)] 名 举止；方式

funeral ['fjuːnərəl] 名 葬礼

arrangement [ə'reɪndʒmənt] 名 布置；安排

reserve [rɪ'zɜːv] 动 保留；预订

thoughtful ['θɔːtfl] 形 深思的；体贴的

respect [rɪ'spekt] 名 尊敬；敬重

upright ['ʌpraɪt] 副 垂直地

pick up 短 捡起；收集

entrance ['entrəns] 名 入口

portion ['pɔːʃn] 名 份

customary ['kʌstəməri] 形 习惯的；惯例的

Practice 请选择合适的词填在下方的横线上

> manners   pick up   entrance   thoughtful   respect

1. Using chopsticks correctly shows _____ for the food and the dining tradition.

2. Chinese table _____ include using chopsticks correctly, like not pointing them at others.

3. The seat facing the _____ is reserved for the most honoured guest.

4. Practice using your chopsticks to _____ food neatly, without dropping it or making a mess.

5. It's _____ to ask if you can have more of a dish before reaching for it.

中国的餐桌礼仪：吃饭的文化 165

Talking practice　情景对话模拟练习

托德很注重礼仪，所以他刚到中国就请教好友杨迪关于中餐的礼仪事项，让我们跟着这段话进行练习吧。

Yang Di, I just arrived in China. Can you tell me what table manners I should follow when eating Chinese food?
杨迪，我刚来中国，请你告诉我，吃中餐有哪些礼仪需要注意？

Todd托德

Yang Di杨迪
Sure, Todd! First, use chopsticks properly. Don't stick them upright in your rice bowl.
当然可以，托德！首先，要正确使用筷子。不要把它们竖插在饭碗里。

Why is that?
为什么呢？

Todd托德

Yang Di杨迪
Because many people think it's not very lucky.
因为很多人认为那样做不太吉利。

Oh, I see. What else should I remember?
哦，我明白了。还有什么需要注意的吗？

Todd托德

Yang Di杨迪
Serve others before yourself. Use serving spoons or chopsticks to put food on others' plates first.
先给别人夹菜。用公筷或公勺先给别人夹菜。

That's very thoughtful! Anything else?
真是体贴！还有吗？

Todd托德

Part 4 独具特色的中国饮食文化

Yes, wait for the host or the eldest person to start eating before you begin.
是的，等主人或长者先动筷子，然后你再开始吃。

Yang Di杨迪

Got it. Thanks, Yang Di! I'll remember these tips.
知道了，谢谢你，杨迪！我会记住这些礼仪的。

Todd托德

Funny facts 关于中国餐桌礼仪的有趣事实和短语

座次讲究：在正式的中餐宴席上，座次是有讲究的。主宾通常坐在面对门口的位置，主人坐在主宾的对面，其他客人按尊卑次序入座。

吃饭速度：中餐讲究细嚼慢咽，吃饭时不要狼吞虎咽。用餐过程应该是放松和享受的时间，快速吃完会被视为不礼貌。

长者先动筷：用餐时，通常由长者或主宾先动筷，其他人才开始用餐。这表示对长者或主宾的尊重。

avoiding waste – 避免浪费
elders first – 长者先动筷
not being picky – 避免挑食

speaking softly – 说话轻声
avoiding bowl tapping – 避免敲碗

Writing practice 写作小练习

根据我们这一节所学到的内容，写出以下句子的英文。

1. 在中国的文化中，餐桌礼仪非常重要。

   _____

2. 吃饭时，不要把筷子插到米饭里。

   _____

3. 等长辈或主人开始吃饭后再吃是一种礼貌。

   _____

Reference translation 参考译文

中国餐桌礼仪是中国文化的重要组成部分。与他人共餐,不仅是品尝美食,更是展现尊重和礼貌的时刻。让我们来了解一些有趣的中国餐桌礼仪吧!

在中国,人们通常用筷子来夹取食物。使用筷子时,有一些规则需要记住。例如,不要将筷子直立插在一碗米饭中,因为这看起来像祭祀时的香,这是不吉利的。另外,不要用筷子指人或玩弄筷子。用筷子轻轻夹起食物放入口中。

在中国的餐桌上,座次安排非常重要。通常,面向入口的座位是留给最尊贵的客人或年长者的。年轻人或地位较低的人则坐在离入口较近的位置。这体现了对长辈和重要客人的尊重。

在中国,先为他人服务再为自己服务是一种礼貌。这意味着在为自己取食之前,你应该使用公筷或公勺为他人夹取食物。这体现了你的体贴和对他人的关心。

中餐往往以"家庭式"方式呈上,将大盘菜肴置于餐桌中央,供众人分享。这种方式使得每个人都能品尝到各种菜肴。重要的是,只取自己能吃得下的分量,避免浪费食物。分享食物也象征着团结与和谐。

和别人一起用餐时,等待主人或年长者先动筷是礼貌的表现。这显示了尊重和良好的礼仪。在大家开始用餐之前,说上一句"开饭了"或"请慢用"也是一种习俗。

中国餐桌礼仪关乎尊重、礼貌和共同享受用餐时光。遵循这些简单的规则,你便能表达出对食物和同伴的珍视。下次用餐时,记住这些中国餐桌礼仪,享受你的用餐体验吧!

Chinese Family Gatherings
中国的家庭聚餐：重要的欢聚时刻

Listening & practice 听英文原声，完成练习

▶扫码听音频◀

1. What special dishes are served at the "one-month-old" party?
   A. Beijing roast duck
   B. Red eggs and ginger
   C. Dumplings and rice cakes

2. What do the bride and groom (新娘和新郎) wear on their wedding day besides Western-style dresses and suits?
   A. Blue clothing
   B. Green clothing
   C. Traditional red clothing

3. What does the fish symbolize during the Chinese New Year reunion dinner?
   A. Health
   B. Abundance (富足)
   C. Happiness

4. What sweet treat is commonly enjoyed during the Mid-Autumn Festival?
   A. Rice cakes
   B. Mooncakes
   C. Dumplings

5. How long do Chinese New Year celebrations usually last?
   A. 10 days
   B. 15 days
   C. 20 days

Reading 阅读下面的文章

China is a country that values family **relationships**, making family gatherings an **essential** part of many people's lives. Whether it's celebrating a baby's one-month-old ceremony, a **wedding**, or traditional holidays, these moments are filled with joy, delicious food, and the warmth of family and friends.

For example, when a baby reaches one month old, Chinese families hold a "one-month-old" party to celebrate this important **milestone**. This tradition dates back centuries and is a way to **introduce** the **newborn** to family and friends. The family hosts a big feast, serving special dishes like red eggs and ginger, symbolizing good luck and protection. Guests bring gifts for the baby and share in the joyful **atmosphere**. It's a time for everyone to share in the happiness of the new arrival.

Chinese weddings are like grand celebrations. On the wedding day, the bride and groom wear not only Western-style wedding dresses and suits but also traditional red clothing that symbolizes happiness and prosperity. The wedding banquet is a highlight, featuring multiple courses of delicious dishes. Each dish has a special meaning, wishing the couple a happy and **prosperous** life together. The celebration includes toasts, laughter, and dancing, making it a memorable event for everyone.

Chinese New Year, also known as the Spring Festival, is the most important holiday in China. On Chinese New Year's Eve, families gather to share a **grand** reunion dinner. The table is filled with symbolic dishes such as dumplings, fish, and rice cakes. Dumplings represent wealth, fish symbolizes **abundance**, and rice cakes signify growth and progress. New Year celebrations usually last for 15 days, ending with the Lantern Festival, where people enjoy lantern displays and eat sweet glutinous rice balls.

The Mid-Autumn Festival is another time for families to gather and admire the full moon. On the 15th day of the eighth lunar month, families come together to share a meal and mooncakes, which are round pastries filled with sweet fillings like lotus seed paste or red bean paste. The round shape of the mooncakes symbolizes family unity and **completeness**. People also light lanterns and share stories under the bright moon. This festival is all about togetherness and expressing love and gratitude.

Chinese gatherings are rich and diverse, bringing people together to celebrate life's important moments. Whether it's a baby's first month, a wedding, or a traditional holiday, these occasions are marked by **joyous** celebrations, delicious food, and the love of family and friends.

Vocabulary and phrases 词汇和短语

relationship [rɪˈleɪʃnʃɪp] 名 关系

wedding [ˈwedɪŋ] 名 婚礼

introduce [ˌɪntrəˈdjuːs] 动 介绍

atmosphere [ˈætməsfɪə(r)] 名 气氛

grand [ɡrænd] 形 豪华的；宏伟的

completeness [kəmˈpliːtnəs] 名 完整；圆满

essential [ɪˈsenʃl] 形 必要的；重要的

milestone [ˈmaɪlstəʊn] 名 里程碑

newborn [ˈnjuːbɔːn] 名 新生儿

prosperous [ˈprɒspərəs] 形 繁荣的；兴旺的

abundance [əˈbʌndəns] 名 丰富；充裕

joyous [ˈdʒɔɪəs] 形 充满快乐的；使人高兴的

Practice 请选择合适的词填在下方的横线上

> relationships   wedding   introduced   newborn   grand

1. Eating together deepens family members _____.

2. Hosting a "one-month-old" party is a way to introduce the _____ to family and friends.

3. At a _____, two families come together to celebrate the happy union of a couple in love.

4. A _____ family gathering brings together many relatives from far and near to celebrate together.

5. During our family dinner, we _____ our cousin who just moved back from abroad.

Talking practice 情景对话模拟练习

凯丽想约陈茹周六去打球，但是陈茹要去参加聚餐，让我们一起来练习这段对话吧。

Chen Ru, are you free to play basketball this Saturday?
陈茹，这个周六你有空一起去打球吗？

Kelly凯丽

Chen Ru陈茹
Sorry, Kelly, I have to go to my nephew's one-month-old party.
对不起，凯丽，这个周六我要去给我的小外甥过满月。

That sounds fun! What do you do at the party?
听起来很好玩！你们会做什么？

Kelly凯丽

Chen Ru陈茹
We have a big feast and wish the baby good health.
我们会举行一个盛大的聚餐，并祝愿小宝宝健康成长。

> What other family gatherings do you have in China?
> 在中国还有哪些家庭聚会?

Kelly凯丽

Chen Ru陈茹

> We also celebrate weddings, Chinese New Year, and the Mid-Autumn Festival.
> 我们还会聚餐庆祝婚礼、春节和中秋节。

Funny facts 关于中国家庭聚餐的有趣事实和短语

团圆饭：家庭聚餐通常称为"团圆饭"，特别是在春节、端午节、中秋节等传统节日中，这象征着家庭团聚和美满。

共同用餐：中国家庭聚餐时，所有菜肴都会摆在桌子中央，大家共同分享，这体现了分享和团结的精神。

丰富多样：家庭聚餐的菜肴通常种类丰富，包括凉菜、热菜、汤、主食和甜点，力求色香味俱全。

- **reunion dinner** – 团圆饭
- **shared dishes** – 共同用餐
- **variety of dishes** – 丰富多样的菜肴
- **celebration rituals** – 庆祝仪式

Writing practice 写作小练习

根据我们这一节所学到的内容，写出以下句子的英文。

1. 家庭聚餐是很多中国家庭的传统习惯。

2. 这个周末，我要去参加外甥的满月酒。

3. 在中秋节聚餐时，我们吃月饼并欣赏美丽的满月。

Reference translation 参考译文

中国是一个重视家庭关系的国家,所以家庭的聚餐,成为了很多人生活中不可或缺的一部分。无论是婴儿满月、结婚,还是传统节日,这些时刻都充满了欢乐、美食和亲友的温暖。

比如当婴儿满月时,中国家庭会举行"满月"宴来庆祝这一重要里程碑。这一传统可以追溯到几个世纪前,是向家人和朋友介绍新生儿的方式。主人家会举办一场盛大的宴会,提供诸如红鸡蛋和生姜等特殊菜肴,象征着好运和保护。客人们带着礼物来庆祝新生儿的到来,分享欢乐的氛围。这是一个让大家共同分享新生命带来的喜悦的时刻。

而中式婚礼犹如盛大的庆典。在婚礼当天,新郎和新娘除了穿着西式的婚纱和礼服以外,还会穿着象征幸福和繁荣的传统红色服装。而婚宴是婚礼的亮点之一,包含多道美味佳肴。每道菜都有特别的寓意,祝愿新婚夫妇幸福美满,生活富足。庆典包括祝酒、欢笑和舞蹈,使每个人都难以忘怀。

中国新年,也被称为春节,是中国最重要的节日。除夕夜,家人们聚在一起,共享盛大的年夜饭。餐桌上摆满了象征意义的菜肴,如饺子、鱼和年糕。饺子象征财富,鱼象征富足,年糕象征成长和进步。新年的庆祝活动通常持续15天,以元宵节作为结束,人们在这一天欣赏灯会,吃甜汤圆。

中秋节更是家庭团聚、赏月的时刻。在农历八月十五这一天,家人们聚在一起吃饭并分享月饼,月饼是一种圆形糕点,内有莲蓉或红豆沙等甜馅。月饼的圆形象征着家庭团圆和完整。人们还会点亮灯笼,在明亮的月光下分享故事。这个节日充满了团聚、爱与感恩的氛围。

中国的聚餐丰富多彩,将人们聚在一起庆祝生活中的重要时刻。无论是婴儿满月、婚礼,还是传统节日,这些场合都充满了欢乐的庆祝、美味的食物和亲友的爱。

Food and Seasons
饮食与节气：顺应自然的饮食之道

Listening & practice 听英文原声，完成练习

▶扫码听音频◀

1. What do people in China eat in spring to help their bodies get ready for summer?
    A. Heavy and rich foods
    B. Fresh and light foods
    C. Spicy and hot foods

2. What is a favourite summer treat in China?
    A. Watermelon
    B. Pumpkin
    C. Hot pot

3. What kind of foods do people eat in autumn to prepare for cooler, dryer months?
    A. Light and fresh foods
    B. Harvest foods like pumpkins and sweet potatoes
    C. Spicy foods

4. What is a popular winter meal in China that helps people stay warm?
    A. Cold noodles
    B. Hot pot
    C. Stir-fried pea shoots

5. What do people in China drink in winter to help warm the body and prevent colds?

A. Green tea

B. Ginger tea

C. Mung bean soup

Reading 阅读下面的文章

In China, people know a lot about how food and health are connected. Chinese food often changes with the seasons. People choose dishes and ingredients that help the body stay healthy with the weather.

Spring is a time of new beginnings. As the weather gets warmer, people in China eat fresh and light foods to help their bodies get ready for summer. They enjoy foods like young leafy greens, bamboo shoots, and peas. These foods help clean the body and get it ready for the active summer months. A popular dish in spring is "stir-fried pea shoots", which are fresh and healthy. Drinking green tea is also common in spring because it is refreshing and helps clean the body.

Summer in China can be very hot and humid, so people eat foods that help them stay cool. They enjoy foods like cucumbers, melons, and mung beans. A favourite summer treat is watermelon, which is refreshing and sweet. People also like soups and drinks made from mung beans because they help cool the body. Another summer favourite is cold noodles, served with vegetables and a light, tangy sauce to keep meals light and cool.

Autumn is the harvest season, and Chinese food in this time helps prepare the body for the cooler, dryer months. People eat foods like pumpkins, sweet potatoes, pears, and lotus roots. These foods keep the body hydrated and nourished. A traditional autumn dish is "lotus root and pork soup", which is believed to help the lungs and boost the immune system. Eating moon cakes during the Mid-Autumn Festival is another special tradition, celebrating family and the harvest.

Winter is cold and dry in many parts of China, so people eat warming and hearty foods to stay warm. They enjoy stews, soups, and dishes made with lamb, beef, ginger, and garlic, which help increase body heat. Hot pot is a popular winter meal where people gather around a simmering pot of broth, cooking meat and vegetables at the table. It's a fun way to stay warm and enjoy a meal with others. Drinking ginger tea is also common in winter because it helps warm the body and **prevent** colds.

In China, eating with the seasons is more than just a tradition; it's a way to stay healthy and in **balance** with nature.

Vocabulary and phrases 词汇和短语

connect [kə'nekt] 动 连接；接通

refreshing [rɪ'freʃɪŋ] 形 新鲜宜人的

mung bean 短 绿豆

lotus root 短 莲藕

immune [ɪ'mjuːn] 形 免疫的

balance ['bæləns] 名 平衡

active ['æktɪv] 形 活跃的；积极的

humid ['hjuːmɪd] 形 潮湿的

prepare [prɪ'peə(r)] 动 预备；准备

nourished ['nʌrɪʃt] 形 滋养的；有营养的

prevent [prɪ'vent] 动 预防；防止

Practice 请选择合适的词填在下方的横线上

connected   active   prepare   prevent   balance

1. After a game of soccer in the fall, we enjoy a hearty stew made with seasonal vegetables to keep us _____.

2. We _____ delicious seasonal fruits like watermelon in summer.

3. Drinking hot soup can help keep us warm and _____ colds in the winter.

4. Chinese food is _____ with the seasons.

5. We adapt our diets to the changing seasons to maintain _____.

Talking practice　情景对话模拟练习

马高向托尼介绍了中国的饮食养生之道，让我们跟着这段对话进行练习吧。

Mao Gao, what are the eating habits of Chinese people?
马高，中国人的饮食习惯都有哪些呢？

Tony托尼

Mao Gao马高

We pay a lot of attention to health. So, our eating habits are closely related to staying healthy. In spring, we eat fresh and light foods like young leafy greens and bamboo shoots to prepare for summer.
我们很注重养生。所以在饮食上，也有很多习惯和健康有关。在春天，我们吃一些新鲜清淡的食物，比如嫩叶蔬菜和竹笋，为夏天做好准备。

What about summer?
那夏天呢？

Tony托尼

Mao Gao马高

In summer, we eat foods like cucumbers and watermelon to stay cool.
夏天，我们吃黄瓜和西瓜等食物来保持凉爽。

And in autumn?
那秋天呢？

Tony托尼

Mao Gao马高

In autumn, we eat pumpkins and sweet potatoes to keep the body hydrated and nourished.
秋天，我们吃南瓜和红薯等食物，保持身体水分和营养。

Winter must be different, too?
冬天肯定也不一样吧?

Tony托尼

Mao Gao马高

Yes, in winter we eat warming foods like stews and hot pot to stay warm.
是的,冬天我们吃炖菜和火锅等热乎乎的食物来保暖。

Funny facts　关于饮食与节气的有趣事实和短语

二十四节气:中国的二十四节气起源于古代农业社会,用来指导农事活动和饮食习惯。每个节气都有特定的饮食习俗,反映了中国人对自然和季节变化的尊重和适应。

立秋:立秋标志着秋天的到来。人们在这一天吃瓜果,以清热润燥,还会吃炖肉等滋补食品,为即将到来的秋冬季节储存能量。

冬至:冬至是二十四节气中非常重要的一个节气,标志着冬天的正式开始。北方人常吃饺子,南方人则吃汤圆,以示团圆和温暖。

Beginning of Spring – 立春	**Beginning of Winter** – 立冬
Beginning of Summer – 立夏	**Summer Solstice** – 夏至
Beginning of Autumn – 立秋	**Winter Solstice** – 冬至

Writing practice　写作小练习

根据我们这一节所学到的内容,写出以下句子的英文。

1. 春天,多吃蔬菜能帮助我们保持健康和活力。

   _____

2. 冬天,我们要吃温暖的食物,如汤和火锅。

   _____

3. 夏天,是时候吃清淡和凉爽的食物来预防中暑(heatstroke)了!

   _____

Reference translation 参考译文

在中国，人们深知饮食与健康的关系。中国的饮食常随季节变化而变化，人们会选择有助于身体适应天气的菜肴和食材。

春天是万物复苏的季节。随着天气变暖，中国人喜欢吃一些新鲜清淡的食物，以帮助身体为夏天做好准备。他们喜欢吃嫩叶蔬菜、竹笋和豌豆等。这些食物有助于清洁身体，为活跃的夏季做好准备。春季的一道受欢迎的菜肴是"清炒豆苗"，这道菜既新鲜又健康。此外，春天喝绿茶也很常见，因为它清爽且有助于排毒。

中国的夏天非常炎热潮湿，所以人们会吃有助于降温的食物。他们喜欢吃黄瓜、西瓜和绿豆等。西瓜是夏天最受欢迎的水果之一，清爽甘甜。人们还喜欢喝绿豆汤和绿豆饮品，因为它们有助于降温。另一个夏季的最爱是"冷面"，搭配各种蔬菜和清爽的酱汁，使餐点清淡又凉爽。

秋天是丰收的季节，中国人在这个季节吃的食物有助于身体适应凉爽干燥的天气。他们喜欢吃南瓜、红薯、梨和莲藕等食物，这些食物有助于保持身体的水分和营养。传统的秋季菜肴有"莲藕猪肉汤"，这道菜被认为有助于润肺，提高免疫力。此外，在中秋节期间吃月饼也是一项特别的传统，人们以此来庆祝家庭和丰收。

中国许多地区的冬季寒冷干燥，所以人们会吃温热且丰盛的食物来保暖。他们喜欢吃炖菜、喝汤并享用以羊肉、牛肉、姜和大蒜做成的菜肴，这些食材有助于增加身体热量。火锅是冬季受欢迎的餐点，人们围坐在煮沸的锅旁，在桌上烹饪肉类和蔬菜。这是一种既有趣又能保暖的聚餐方式。冬天喝姜茶也很常见，因为它有助于暖身和预防感冒。

在中国，顺应季节变化饮食不仅仅是一种传统，它还是保持健康和与自然和谐相处的方式。

参考答案

Part 1 中国美食的代表菜肴

Beijing Roast Duck 北京烤鸭：北京菜的经典代表

Listening & Practice 听英文原声，完成练习
1. C   2. C   3. B   4. B   5. C

Practice 请选择合适的词填在下方的横线上
1. process   2. unique   3. crispy   4. recipe   5. texture

Writing Practice 写作小练习
1. I like eating roast duck, its skin is crispy, and the meat is tender.
2. Beijing roast duck has a long history.
3. Beijing roast duck is one of the most famous specialties in Beijing.

Sichuan Hot Pot 四川火锅：真正的"热辣滚烫"

Listening & Practice 听英文原声，完成练习
1. C   2. A   3. A   4. C   5. B

Practice 请选择合适的词填在下方的横线上
1. flavour   2. gather   3. experience   4. chili peppers   5. a variety of

Writing Practice 写作小练习
1. People in Sichuan like to eat hot pot and add a lot of spicy chili peppers.
2. You can cook many kinds of food in a Sichuan Hot Pot, such as meat, vegetables, tofu, and noodles.
3. Eating Sichuan Hot Pot with family and friends is really fun!

Kung Pao Chicken 宫保鸡丁：历史悠久的菜肴

Listening & Practice 听英文原声，完成练习
1. B   2. B   3. A   4. B   5. C

Practice 请选择合适的词填在下方的横线上
1. tasty   2. protein   3. imagine   4. healthy   5. create

Writing Practice 写作小练习
1. In Kung Pao Chicken, you can find chicken, peanuts, and colourful vegetables.
2. Kung Pao Chicken is a famous dish from Sichuan!
3. Eating Kung Pao Chicken with rice is a wonderful way to enjoy this delicious food.

Sauteed Shrimp with Longjing Tea 龙井虾仁：西湖边上的美味

Listening & Practice 听英文原声，完成练习
1. B   2. C   3. B   4. B   5. A

Practice 请选择合适的词填在下方的横线上
1. shrimp   2. favour   3. showcases   4. cuisine   5. nutritious

Writing Practice 写作小练习
1. Sauteed Shrimp with Longjing Tea is not only delicious but also very healthy.
2. It is made with tasty shrimp and Longjing tea.
3. Sauteed Shrimp with Longjing Tea is a special Chinese dish.

Dongpo Pork 东坡肉：以苏东坡命名的经典菜肴

Listening & Practice 听英文原声，完成练习
1. A   2. C   3. B   4. B   5. C

Practice 请选择合适的词填在下方的横线上
1. squares   2. flavourful   3. significant   4. talented   5. layer

Writing Practice 写作小练习
1. Dongpo Pork is made with pork belly and a special sauce.
2. The name Dongpo Pork comes from the great Song Dynasty writer Su Dongpo.
3. It needs to be cooked slowly to make the meat very tender.

Crab Meat Lion's Head 蟹粉狮子头：扬州的传统名菜

Listening & Practice 听英文原声，完成练习
1. B   2. A   3. C   4. C   5. B

Practice 请选择合适的词填在下方的横线上
1. meatball   2. gently   3. broth   4. enjoyable   5. originated

Writing Practice 写作小练习
1. Crab Meat Lion's Head is made with crab meat and pork, shaped like a lion's head.
2. Mix the crab meat and pork together, then form them into large meatballs.
3. Crab Meat Lion's Head is cooked in a delicious broth.

Buddha Jumps Over the Wall 佛跳墙：浓郁醇香的享受

Listening & Practice 听英文原声，完成练习
1. B   2. C   3. B   4. A   5. A

Practice 请选择合适的词填在下方的横线上
1. origin   2. return   3. jumping over
4. contains   5. chef

参考答案

Writing Practice 写作小练习

1. Buddha Jumps Over the Wall is made with many different ingredients, such as chicken, duck, and seafood.
2. The aroma of Buddha Jumps Over the Wall is so enticing that even Buddha would jump over the wall to taste it.
3. Buddha Jumps Over the Wall is a famous dish from Fuzhou, China, and it is really delicious.

Boiled Chicken Slices 白切鸡：粤菜中的经典

Listening & Practice 听英文原声，完成练习

1. B   2. A   3. B   4. A   5. A

Practice 请选择合适的词填在下方的横线上

1. simple   2. smooth   3. classic   4. Unlike   5. perfectly

Writing Practice 写作小练习

1. Boiled Chicken Slices is a classic dish in Cantonese cuisine.
2. Boiled Chicken Slices is not only delicious but also very healthy.
3. We often eat Boiled Chicken Slices during festivals and family gatherings.

Part 2 丰富多彩的中国地方小吃

Yangrou Paomo 羊肉泡馍：来自西北的温暖

Listening & Practice 听英文原声，完成练习

1. C   2. A   3. B   4. A   5. B

Practice 请选择合适的词填在下方的横线上

1. describe   2. tear   3. flatbread   4. a bit of   5. warming

Writing Practice 写作小练习

1. Yangrou Paomo is a delicious Chinese snack.
2. This snack is warm and comforting, perfect for cold weather.
3. It is made by soaking flatbread in warm lamb broth.

Duck Blood and Vermicelli Soup 鸭血粉丝汤：金陵的独特美味

Listening & Practice 听英文原声，完成练习

1. C   2. B   3. C   4. A   5. B

Practice 请选择合适的词填在下方的横线上

1. accidentally   2. unusual   3. definitely
4. belongs to   5. source

Writing Practice 写作小练习
1. Duck Blood and Vermicelli Soup is a very tasty soup.
2. It contains thin vermicelli and duck blood.
3. I like drinking Duck Blood and Vermicelli Soup because it feels warm.

Harbin Red Sausage 哈尔滨红肠：满口都是肉香

Listening & Practice 听英文原声，完成练习
1. C   2. C   3. B   4. B   5. A

Practice 请选择合适的词填在下方的横线上
1. stew   2. sausage   3. smoky   4. cultural   5. popular

Writing Practice 写作小练习
1. Harbin Red Sausage is a famous food from Harbin, China.
2. It is made with pork and spices, and perfectly smoked.
3. People like to eat Harbin Red Sausage with bread.

Shanghai Soup Dumpling 上海小笼包：充满鲜美的汤汁

Listening & Practice 听英文原声，完成练习
1. B   2. A   3. B   4. B   5. C

Practice 请选择合适的词填在下方的横线上
1. find out   2. melt   3. wonderful   4. dough   5. Throughout

Writing Practice 写作小练习
1. Shanghai Soup Dumplings are a popular snack from Shanghai, China.
2. I like to eat the dumplings with vinegar and ginger sauce.
3. The filling of Shanghai Soup Dumplings is usually made with pork and vegetables.

Guangxi Luosifen 广西螺蛳粉：有点"臭臭的"酸爽

Listening & Practice 听英文原声，完成练习
1. A   2. A   3. B   4. A   5. B

Practice 请选择合适的词填在下方的横线上
1. chewy   2. river snail   3. experiment   4. various   5. unforgettable

Writing Practice 写作小练习
1. Guangxi Luosifen are made with rice noodles, river snails, and various spices and vegetables.
2. Its unique flavour comes from the spicy, tangy, and savory broth.
3. If you like spicy food, don't miss Guangxi Luosifen.

Guangzhou Steamed Vermicelli Rolls 广州肠粉：软糯可口的早餐

Listening & Practice 听英文原声，完成练习
1. B   2. B   3. B   4. A   5. C

Practice 请选择合适的词填在下方的横线上
1. locals   2. absolutely   3. Among   4. rolling up   5. silky

Writing Practice 写作小练习
1. Steamed Vermicelli rolls have a soft and chewy texture.
2. Guangzhou Steamed Vermicelli Rolls are a popular breakfast food in Guangzhou, China.
3. They are made with thin rice flour sheets wrapped around various fillings.

Yunnan Cross-Bridge Rice Noodles 云南过桥米线：独特的米线体验

Listening & Practice 听英文原声，完成练习
1. B   2. A   3. A   4. B   5. B

Practice 请选择合适的词填在下方的横线上
1. separate   2. remains   3. heritage   4. delightful   5. method

Writing Practice 写作小练习
1. Cross-Bridge Rice Noodles are made by adding the rice noodles and the raw ingredients to the steaming broth.
2. You can add ingredients like vegetables, meat, and eggs according to your preference.
3. Eating Cross-Bridge Rice Noodles is an enjoyable experience.

Tianjin Jianbing Guozi 天津煎饼馃子：天津的经典早餐

Listening & Practice 听英文原声，完成练习
1. C   2. B   3. B   4. A   5. B

Practice 请选择合适的词填在下方的横线上
1. thus   2. skillfully   3. invented   4. includes   5. commercial

Writing Practice 写作小练习
1. Jianbing Guozi is a popular street snack in China.
2. It is made by spreading a thin layer of batter on a hot griddle.
3. Jianbing Guozi is usually topped with savory sauce and sometimes an egg is added.

Lanzhou Beef Noodles 兰州牛肉面：兰州的特色美食

Listening & Practice 听英文原声，完成练习
1. A   2. B   3. C   4. A   5. A

Practice 请选择合适的词填在下方的横线上
1. entire   2. requires   3. offers
4. element   5. generation; generation

Writing Practice 写作小练习
1. Lanzhou Beef Noodles are made with hand-pulled noodles.
2. The broth is rich and flavourful, often made by simmering beef bones.

3. Lanzhou Beef Noodles also include tender slices of beef and fresh vegetables.

Xinjiang Lamb Skewers 新疆羊肉串：香气四溢的烧烤

Listening & Practice 听英文原声，完成练习
1. C   2. B   3. A   4. B   5. B

Practice 请选择合适的词填在下方的横线上
1. powder   2. tradition   3. lamb   4. centuries   5. flame

Writing Practice 写作小练习
1. Xinjiang not only has beautiful scenery but also delicious grilled lamb skewers.
2. The lamb is grilled over high-temperature coal, giving it a smoky and flavourful taste.
3. Tasting Xinjiang Lamb Skewers is a wonderful way to enjoy the flavours of Xinjiang cuisine.

Part 3 历史悠久的中国节日美食

Mooncake 月饼：中秋节的赏月佳品

Listening & Practice 听英文原声，完成练习
1. C   2. B   3. B   4. A   5. B

Practice 请选择合适的词填在下方的横线上
1. meaningful   2. admire   3. ancient   4. mysterious   5. symbolizes

Writing Practice 写作小练习
1. Mooncakes have various flavours, such as lotus seed paste and red bean paste.
2. Mooncakes are delicious round pastries commonly eaten during the Mid-Autumn Festival.
3. During the Mid-Autumn Festival, people give mooncakes to friends and family as gifts.

Zongzi 粽子：拥有 2000 年历史的美食

Listening & Practice 听英文原声，完成练习
1. A   2. C   3. B   4. C   5. B

Practice 请选择合适的词填在下方的横线上
1. pyramids   2. glutinous   3. tightly   4. minister   5. express

Writing Practice 写作小练习
1. Zongzi are made with glutinous rice wrapped in bamboo leaves.
2. People eat Zongzi during the Dragon Boat Festival to commemorate the poet Qu Yuan.

3. My mom often cooks Zongzi for me during the Dragon Boat Festival.

Yuanxiao 元宵：象征团圆的小吃

Listening & Practice 听英文原声，完成练习
1. B 2. A 3. B 4. C 5. A

Practice 请选择合适的词填在下方的横线上
1. reunion 2. provides 3. dessert 4. riddles 5. wonderfully

Writing Practice 写作小练习
1. Yuanxiao is a dessert we enjoy during the Lantern Festival.
2. The round shape of yuanxiao symbolizes happiness and reunion.
3. I like eating yuanxiao because it is soft, chewy, and very sweet.

Dumplings 饺子：有"年味儿"的美食

Listening & Practice 听英文原声，完成练习
1. B 2. C 3. A 4. A 5. B

Practice 请选择合适的词填在下方的横线上
1. excellent 2. ward off 3. whoever 4. contribute 5. Freshly

Writing Practice 写作小练习
1. Making dumplings is very fun.
2. The filling of dumplings can be meat, vegetables, or seafood.
3. Dumplings are shaped like little boats.

Laba Porridge 腊八粥：腊八节的传统美食

Listening & Practice 听英文原声，完成练习
1. A 2. B 3. C 4. A 5. C

Practice 请选择合适的词填在下方的横线上
1. Porridge 2. community 3. typically 4. satisfying 5. approaches

Writing Practice 写作小练习
1. Laba Porridge keeps us warm and nourished during the cold winter.
2. Laba Porridge is made with various grains, beans, nuts, and dried fruits.
3. Laba Porridge tastes sweet and delicious.

Nian gao 年糕：新年步步高升的象征

Listening & Practice 听英文原声，完成练习
1. A 2. A 3. A 4. A 5. B

Practice 请选择合适的词填在下方的横线上
1. mentioned 2. common 3. endless 4. accompanies 5. ensure

Writing Practice 写作小练习

1. niangao is made with glutinous rice flour, giving it a sticky and chewy texture.
2. Eating niangao symbolizes good luck and prosperity in the New Year.
3. People often add red beans, nuts, or other ingredients to niangao.

Part 4 独具特色的中国饮食文化

Regional Food Cultures in China 中国的地方饮食文化：风味之旅

Listening & Practice 听英文原声，完成练习
1. C   2. C   3. B   4. C   5. A

Practice 请选择合适的词填在下方的横线上
1. numbing   2. Influenced   3. appreciate   4. geography   5. ethnic

Writing Practice 写作小练习

1. People in Guangdong like to have morning tea and taste various dim sum.
2. Different places in China have their own unique foods.
3. In China, food culture represents tradition and customs.

Chinese Cooking Techniques 中国的烹饪技巧：中餐美味的秘诀

Listening & Practice 听英文原声，完成练习
1. A   2. B   3. C   4. C   5. A

Practice 请选择合适的词填在下方的横线上
1. develop   2. stacked   3. bring out   4. technique   5. varied

Writing Practice 写作小练习

1. Steaming is a healthy cooking method that preserves the natural flavours of the food.
2. Boiling is very simple; you just need to cook the food in water until it is done.
3. Roasting makes the outside of the food crispy while keeping the inside soft and juicy.

Chinese Table Manners 中国的餐桌礼仪：吃饭的文化

Listening & Practice 听英文原声，完成练习
1. A   2. A   3. B   4. A   5. A

Practice 请选择合适的词填在下方的横线上
1. respect   2. manners   3. entrance   4. pick up   5. thoughtful

Writing Practice 写作小练习
1. In Chinese culture, table manners are very important.
2. When eating, do not stick your chopsticks into the rice.
3. It is polite to wait until the elders or the host start eating before you begin.

Chinese Family Gatherings 中国的家庭聚餐：重要的欢聚时刻

Listening & Practice 听英文原声，完成练习
1. B   2. C   3. B   4. B   5. C

Practice 请选择合适的词填在下方的横线上
1. relationships   2. newborn   3. wedding
4. grand   5. introduced

Writing Practice 写作小练习
1. Family gatherings for meals are a traditional habit for many Chinese families.
2. This weekend, I am going to attend my nephew's one-month-old ceremony.
3. During the Mid-Autumn Festival dinner, we eat mooncakes and admire the beautiful full moon.

Food and Seasons 饮食与节气：顺应自然的饮食之道

Listening & Practice 听英文原声，完成练习
1. B   2. A   3. B   4. B   5. B

Practice 请选择合适的词填在下方的横线上
1. active   2. prepare   3. prevent   4. connected   5. balance

Writing Practice 写作小练习
1. In spring, eating more vegetables can help us stay healthy and energetic.
2. In winter, we should eat warm foods like soup and hot pot.
3. In summer, it's time to eat light and cool foods to prevent heatstroke!